Nature's Second Kingdom

Marcello Malpighi, "The Anatomy of Plants," frontispiece to *Opera omnia* (London, 1687). Photo courtesy of the Muséum national d'histoire naturelle, Paris.

Nature's Second Kingdom

Explorations of Vegetality in the Eighteenth Century

François Delaporte

translated by

Arthur Goldhammer

The MIT Press

Cambridge, Massachusetts

London, England

English translation © 1982 by the Massachusetts Institute of Technology

Originally published in France under the title *Le Second Règne de la nature*, copyright © 1979 by Flammarion, Paris

All rights reserved. No part of this book may be reproduced in any form or by any means, electronic or mechanical, including photocopying, recording, or by any information storage and retrieval system, without permission in writing from the publisher.

This book was set in Sabon by Graphic Composition, Inc. and was printed in the United States of America.

Library of Congress Cataloging in Publication Data

Delaporte, François, 1941–
   Nature's second kingdom.

  Translation of: Le second règne de la nature.
  Bibliography: p.
  Includes index.
  1. Botany—History—18th century. I. Title.
QK15.D4413   581'.09'033   81–19338
ISBN 0–262–04066–2   AACR2

Contents

	Translator's Note	vii
	Foreword by Georges Canguilhem	ix
	Introduction	1
1	*The Central Problems of Botany*	9
2	*Nutrition*	29
3	*Generation*	91
4	*Movement*	149
	Conclusion	187
	Appendix	193
	Notes	199
	Bibliography	233

Biographical Glossary 253

Name Index 263

Translator's Note

François Delaporte has made extensive use in this work of sixteenth-, seventeenth-, and eighteenth-century texts in a number of languages. When a work written in English is cited, or when an English translation from the period has been available, I have tried to find the passage in the original, rather than retranslate from the author's French rendering. Since many of the works cited are quite rare, however, this has not always proved possible. I have tried to indicate in the notes wherever multiple translation, with its concomitant increased likelihood of distortion, is involved. Italics in all quotations are as in the original.

Foreword

Plant is a word that remained in common usage at a time when the word *vegetable* came to be restricted to scientific parlance. The sedentary character of agricultural societies, which draw their sustenance from the soil, was based on the immobility of plants, just as the nomadic character of pastoral societies was based on the mobility of animals. Compared with knowledge of animal life, knowledge of plant life long remained as immobile as plants themselves. Only after the first glimmerings of knowledge of vital functions in animals had been acquired through analytical methods of experiment did progress begin to be made in knowledge of plants. Knowledge of animal functions preceded knowledge of plant functions, or *vegetality*, for one very good reason: the seekers of knowledge were animals. In studying animals, men sought a substitute for studying their own nature. The verb *to vegetate*, originally derived from Latin words meaning strength and growth, quickly came to signify inertia and apathy. How could plants be expected to enlighten mankind as to what makes men active, curious, and conquering creatures? Thus, in his *History of Botany*, Julius von Sachs was able to write, "Little more was known of vegetable life during the sixteenth and early seventeenth centuries than had been known since the earliest times of civilization, and what little was known stemmed from the practice of agriculture, gardening, and, more generally, from practical occupations related to plants."

Beyond any doubt, knowledge of plant structures and functions received a decided lift from the work of eighteenth-century botanists. Every history of botany makes the point. Most of these histories, however, are a long way from asking why and under what circumstances the naturalists of this period worked out their tactics and methods. To ask such questions in the history of a science, one must take an interest not so much in its results, celebrated as victories, as in the way problems were formulated, even if they remained unresolved. Science must be seen as a painstaking effort to interpret phenomena, using certain hypotheses as a key. Accordingly, the genesis of these hypotheses must take precedence over the classification of observations. One object of the present work by François Delaporte is to put certain epistemological concepts such as analogy and model to the test in order to gauge how fruitful or unfruitful specific heuristic choices turned out to be. Shrewdly, though, Delaporte aims to do more than this. Analogies and models do not appear by themselves. They are chosen. The inspiration behind such choices reveals the latent presence of paradigmatic values and schemes of collective apperception characteristic of a particular place, time, and culture.

Naturalists working on plant nutrition who thought they had been able to demonstrate that the sap of plants circulates were of course strongly influenced by the growing renown of Harvey's discovery, but they were also succumbing to a more general tendency to take a unitary view of natural forces that grew out of the work of Newton and Descartes.

The controversies over sexuality in plants and over the ways and means of fertilization were not entirely innocent of fantasies concerning amorous behavior, which some found enticing and others repugnant. Botany, morality, and religion passed arguments back and forth.

Finally, the study of plant movement was just as important as the study of reproduction in locating the plant kingdom relative to the other two kingdoms in nature's hierarchy, the animal and the mineral. In animals, it was held, movement is the manifestation of instinct, need, and feeling. Movement reflects the necessity to adapt to the environment and to changing conditions. Do comparable adaptive functions exist in plants? And are these functions subordinate to some ultimate purpose? Here, too, botany and theology found common ground.

In the eighteenth century, questions pertaining to vegetality were ultimately questions of phytophilosophy as much as phytophysiology. In the final pages of his essay, François Delaporte gives a brief but clear account of the break in continuity between the work of eighteenth-century naturalists and early nineteenth-century biologists, a break that marks the beginning of progress that has continued without interruption to the present day.

Readers will surely take much pleasure from this work; the author displays great erudition and yet never lapses into pedantry; he has been able to show how one way of investigating plant life was essentially outmoded and yet produced results that subsequent knowledge has not made totally obsolete. This ambivalence is characteristic of eighteenth-century knowledge. Jean-Jacques Rousseau begins the article "Flowers" in his *Fragment pour un dictionnaire des termes d'usage en botanique* with the following words: "If I surrendered my imagination to the sweet sensations that this word seems to evoke, I could write an article that might be very agreeable to shepherds but very bad for botanists: for a moment, then, let us forget about lively colors, sweet fragrances, and elegant shapes, and first try to obtain a good understanding of the organized being in which they are united." It is easier to renounce the fashion for the pastoral style than to silence the shepherd in one's soul. Before becoming the subject matter of theory, trees, flowers, fruits, and grains are symbols, spurs to the imagination. Plant symbolism continued to haunt the deliberate attempts of eighteenth-century botanists to depict the plant as a mechanism, an economy, or an organism. Rousseau's contemporary Goethe, a poet who sought in the study of plants something more than Rousseau was looking for, namely, a principle of natural philosophy, and who found himself irritated by the lukewarm reception accorded his *Essay on the Metamorphosis of Plants*, wrote that "no one wanted to understand the intimate union of poetry and science; it was forgotten that poetry is the source of science." Indeed, before flowers could become simply objects of science, men had to learn to see them, in the words of Mallarmé, as "missing from every bouquet."

Georges Canguilhem

Nature's Second Kingdom

Introduction

In the history of vegetable physiology, the history of animal physiology has generally served to provide a model of intelligibility. Historians find that this area of botany was similar in character to seventeenth-century medicine. Just as human anatomy was inextricably bound up with physiology, so physiological observation of the vegetal functions was inseparable from the study of organs. In consequence it has been maintained that first the parts of plants had to be identified before their use could be examined; the study of vegetal functions, we are told, presupposes the existence of a phytotomy. This has led some authors to stress the supposed importance of refined techniques of dissection and, above all, the use of the microscope.

Plant physiology, then, is supposed to have been subject to the same vicissitudes as animal physiology. Accordingly mechanism is said to have dominated the subject until the end of the seventeenth century, at which point, so we are told, most botanists were forced to turn their attention to the animal kingdom, since it had become impossible to explain how the vegetal machine worked in terms of the laws of motion alone. Thus the shift to a new approach, which consisted in explaining the inferior by the superior, the vegetable by the animal, ostensibly signaled the end of seventeenth-century mechanism.

Three versions of this tale may be found in historical accounts of research into vegetal functions. Usually all three are mingled. In the first version, historians focus on the relations between the analogical method and the experimental method. From the first a key role is assigned to the search for similarities between plants and animals. Take nutrition, for example. "Once Harvey had discovered the circulation of the blood, men began to wonder if in plants there might be a circulation of sap. A hypothesis was therefore put forward, an initial problem formulated, the solution of which required observational or, better still, experimental methods."[1] The argument, then, is that it was not until William Harvey had made his discovery that botanists began to suspect the existence of a circulation of sap in plants. Therein lay a primary problem to be resolved by means of observation and experimentation. The only role of analogy was to suggest a hypothesis for testing. On this view, then, what Stephen Hales did was to refute by experiment a theory elaborated thirty years earlier by Daniel Major or Claude Perrault.

This manner of proceeding has its disadvantages. To link events together in this way is to give a linear account of history, taking no notice of accident or discontinuity. But it is also to reconstruct the past in a manner that is historically incorrect, for refuting the theory of the circulation of sap was neither Stephen Hales's primary concern nor even the point of departure of his experiments. And yet another insurmountable difficulty confronts us. It is impossible to explain why the theory of the circulation of sap should have had supporters throughout the eighteenth century, even though the question had been definitively settled by Stephen Hales in 1727.

In the second common version of this history, the accent is placed on the advantages but also on the frustrations inherent in the analogical and experimental methods. In truth the question is one of determining whether these two methods have been used correctly or incorrectly. The historian therefore is led to take into account such subjective factors as the scientist's perspicacity or purblindness, his skill or clumsiness. Take reproduction, for example. The use of analogy, it would seem, is legitimate, even necessary, when it results in the observation and description of the sexes: "The discovery of sexuality in plants could not have come to pass but for comparison of certain phenomena of vegetal life with reproduction in animals."[2] On the other hand, the use of analogy is open to criticism when the theory to which it leads turns out to be erroneous. This is the case with the theory of agamy, according to which plants bear fruit without fertilization. If Lazzaro Spallanzani adopted this theory, Jean Rostand tells us, it is because "Spallanzani must have brought with him to this new field of study [vegetable physiology] his thoroughly ovist and preformationist prejudices."[3] Toward the experimental method we find the same ambivalence. On the one hand, when the focus is on stamen resection experiments, the experimental method is said to have achieved decided progress. It was Rudolph Jacob Camerarius who demonstrated the role of pollen in fertilization: "Only by experiment could vegetable sexuality be proven . . . A start had to be made by proving experimentally that the fertilization of the seed is the work of the pollen."[4] On the other hand, the experimental method sometimes ended in impasse. For evidence one has only to allude to the same experiments of Lazzaro Spallanzani, which now prove the reverse of what they proved a moment ago. This is because, as Jean Rostand says, "if there is no reason to doubt, a priori, that Spallanzani did in fact carry out the experiments he published on the generation of plants, there are nevertheless grounds for believing that they were not conducted with all the rigor and skill required."[5]

To focus, then, on the similarities between animal and vegetable physiology and to protest when those similarities are exaggerated is to hold that the influence of each discipline on the other was sometimes a boon, sometimes a hindrance, to progress, depending on how much light happened to be shed in any particular case. The same remark holds good if the historian focuses on the skill or ineptness of the experimenters. This manner of writing history is not without its defects. Is not psychologism the besetting danger of this type of analysis? Just as the heuristic worth of an analogy depends on the sagacity of the scientist, so the value of an experiment depends on his skill.

In the third common version of this history, historians point to the purely metaphoric, not to say fantastic, significance of the analogies drawn between plants and animals. At the same time they give prominence to the work of observers and experimenters, some of whom are described as precursors working in complete isolation from their contemporaries, while others are hailed as founders of a scientific discipline. Take plant motion, for example. The first step in the historical account is to discredit the analogical method: "Between 1770 and 1815 the subject of plant irritability was in the hands of speculative essayists, who gave the idea of botanical analogy its most complete expression. . . . In regard to vegetable motion, so long as naturalists did not reject all thought of animal analogies, no real progress was made." The next step is to accredit the experimental method: "Not until the experimental researches of John Hunter, John Lindsay, and Thomas Knight would it become possible to refer to any protracted observations on isolated phenomena of irritability and tropisms in plants. Their investigations will be discussed as one part of a revival of experiment in natural history, coincident with the development of the science of biology."[6]

This version of the history of botany, by far the most widely current of the three, seems ultimately to rest on two preconceived notions. The first is that the detection of analogies was a cause of stagnation; the second, that only a method based on observation and experimentation could bring progress in vegetable physiology.[7] These two notions complement each other remarkably well. The analogical method was a hindrance, it is argued, because it was a procedure the eighteenth century presumably inherited from the Renaissance or even Antiquity. The identification of the various parts of the vegetable organism with the known parts of the animal was, we are told, a source of error. In contrast, the work done by experimentalists and observers was supposedly a prefiguration of the nineteenth century. This approach to history has not a few disadvantages of its own. Knowledge has to be divided into two separate, albeit intersecting, domains. One domain used theories and concepts that were transferred from animal physiology to botany (including circulation, preformationism, and irritability), which allegedly resulted in a tissue of errors: the circulation of sap, the contention that plants lack sexuality, and the notion of vegetable irritability. The other domain was ostensibly based on rigorous empirical methods for investigating the vegetable kingdom. Out of these investigations, we are told, came the first laws of vegetation. Sexuality was shown to exist in vegetables. Soon thereafter came a theory of pollination, and finally plant motions were explained for the first time in terms of the laws of physics.

That the discipline has had to be divided up in this way is one indication that something has gone wrong. For one thing, the logical connections among certain statements are obscured, because botanists numbered among the analogists happen to have become involved in observation or experiment. For another, the use of analogy by those described as observers and experimentalists is disguised. In the end, to write history in this way is to argue that those who used analogies that led to what we now see were errors were completely misguided, whereas those who made observations and did experiments were thereby assured of always being squarely on the path of truth. As long as analogies were used, the study of plants, so the argument runs, could not progress beyond the early stages of scientific development, whereas the knowledge obtained during the eighteenth century from observation and experiment needed only to be refined in the following century. As Michel Foucault has pointed out, the main defect of this approach to the history of science is "the application to scientific knowledge of strictly anachronistic categories. 'Life' is obviously the chief among these. Historians want to write the history of biology in the eighteenth century; but they do not realize that biology did not then exist, and that the pattern of knowledge that has been familiar to us for a hundred and fifty years is not valid for a previous period."[8]

If this is true, someone may well object that it is also anachronistic to study the history of "vegetality" in the eighteenth century. If it is not valid to study plant biology, how is it possible to write the history of vegetality? What is meant by vegetality, if not a series of phenomena common to both plants and animals, in the forms in which those phenomena appear in plants? Such phenomena include nutrition, growth, and reproduction.[9] Contractile and sensitive properties, on the contrary, were supposed to be characteristic of animality, as opposed to vegetality. But we shall also be looking at the problem of plant motions, which were studied by comparing them to animal motions. This being the case, how can the claim be justified, that this book is a study of the history of vegetality in the eighteenth century? To begin with, notice that the subject of the study is the historicity of a branch of knowledge concerned with the nature of the vegetable kingdom and all its properties and, therefore, with what people thought about the vegetable kingdom. As it happens, in the eighteenth century the notion of vegetable irritability and sensitivity figured in people's thinking in this regard. And this is not all. Although in the eighteenth century the term *vegetality* appears to have been used only once, it was used in a sense that makes it legitimate to include a historical account of the problem of plant motion in a study of vegetality. "Irritability is a faculty peculiar to animality; several botanists claim, however, to have demonstrated its existence in the realm of vegetality." [10]

The study of the history of vegetality in the eighteenth century is not a preparation for, not even an adumbration of or a preface to, the study of the history of vegetality in the nineteenth century. Our primary objective will be to look at what made it possible to formulate problems of vegetable physiology and to show how the classical age [i.e., the seventeenth and eighteenth centuries] posed those problems. The overall aim is to envision what might be called the "problematic" of botany. We shall then turn to the following three questions: the question of nutrition, the question of reproduction, and the question of plant motion. To describe the various means used by botanists to establish the existence of these functions is to emphasize the importance of reasoning by analogy, but also to show that analogies were a source of conflicting views. I shall therefore point out in what respects the various models and terms of reference used were incompatible with one another. What other hope is there of explaining how botanists came to propose a variety of systems or mechanisms to account for each of the major functions?

Next we shall turn our attention to the different ways in which scientific discourse was elaborated. Even after a model from animal physiology had been imported into the domain of vegetable physiology, the associated function still had to be constituted. Hence botanists immediately looked for structural and functional resemblances between plants and animals. For this they employed a shifting array of quite crude techniques, with the result that vegetal forms and functions were assimilated to those of animals. There was enough flexibility in establishing correspondences, however, that prejudgment of the nature of functional mechanisms in vegetables was avoided. But a question arose: Was it or was it not possible to imagine organs in the plant similar to those found in the animal? To answer this question comparisons had to be made between animals and vegetables. These revealed similarities in some cases, differences in others. Now, however, the techniques of identification employed were more refined and rigorous than had been the case previously.

Finally, we shall consider the status of the various statements constituting a scientific discourse and the different kinds of relations that may be established between one statement and another. One such relation is that between proposition and confirmation. Statements were arranged so that the demonstrative part of the assertion could not but confirm—in the etymological sense of the word—theories implicitly held to be true. The relation of hypothesis to verification is another example. In this case the demonstrative part of the assertion consisted in submitting to experimental control, hence to verification, theories whose veracity was in question. While focusing on these key relations, however, we shall not neglect subsequent systematizations.

With regard to each function, finally, we shall consider two series of questions. First, anyone who explains the vegetable in terms of the animal must deal with problems concerning the similarities and differences between these two classes of beings in relation to the function under consideration. These problems cannot be solved if vegetable functions are assimilated to animal ones. Botanists who see differences in the organs through which the functions are expressed, however, can resolve the difficulty by showing how the functions common to both plants and animals are carried out in different ways in each. Once this question is out of the way, we shall look at the different strategies used to assign statements in the field of vegetable physiology a place and a role in the realm of metaphysics, as well as in that of morals, taking care not to neglect the function of desire in relation to physiological discourse.

I   *The Central Problems of Botany*

What accounts for the interests of naturalists in the nature of plant life? "When one is fond of reflecting on things, it is impossible to gaze constantly on remarkable events taking place before one's very eyes without being curious about their history and how to read it."[1] Moving beyond noble curiosity, Sénebier adds a further explanation, focusing on ways of improving production: "Society's interest calls loudly for studies of this kind, and society's interest should be enough of a motive to pursue them. Such knowledge is not idle, but needed by agriculture: without correct ideas of the organization of plants, agriculture can hardly be improved."[2] At bottom, Sénebier is interested not so much in explaining why botany is studied as in justifying that study. The study of plants without consideration of their uses was scarcely conceivable. Before plant life could become an object of study, an obstacle of technological interest had to be surmounted.

Another benefit was to be hoped for from knowledge of plants too. Such knowledge "might instruct us as to the plan of the universe; it might uncover formulas that hide thousands of secrets of animal economy."[3] The idea of using inferior forms of life to shed light on superior forms was not a new one, having been advanced by Malpighi at the end of the seventeenth century. In reality, the opposite took place. What were the reasons for this reversal? Why was animal life initially taken as a model for the explanation of vegetal functions? And what methods did botanists use in elaborating this explanation? Before we look into these questions, we must ask how the classical age circumscribed the domain of plant anatomy within the broader domain of knowledge of living things. And we must consider another, complementary question: Did this anatomy make it possible to discuss how the plant used its different parts? If it did not, how are we to explain the seemingly paradoxical fact that there emerged a plant physiology which, unlike animal physiology, was independent of anatomy, or at any rate of that type of plant anatomy said to have been founded by Grew and Malpighi?

PLANT ANATOMIES AND PLANT PHYSIOLOGY

Inevitably, man must form some idea of the complexity of plant structure from even the humblest of his arts and the most commonplace of his actitivites. The reason is that to break down plant tissue, we have to take an approach quite unlike that used to pulverize or chip a sample of mineral. For the latter, striking a blow or exerting pressure will suffice, whereas plants, being of less dense composition than minerals, can be bent, compressed, snapped off. The need to cut plants has induced men to devise a bewildering variety of tools, and wielding them is enough to persuade even the humblest worker of the complexity of plant structure. From this fact, however, one might equally well draw the opposite conclusion, namely, that plant structure could not be seen as complex because it did not exist, at least not as an object of study. The more widely plants were used in such practical pursuits as farming, industry, and medicine, the more opportunities there were to censor observations of their nature and the less desire there was to comprehend. Mankind's practical labors did not bring knowledge of

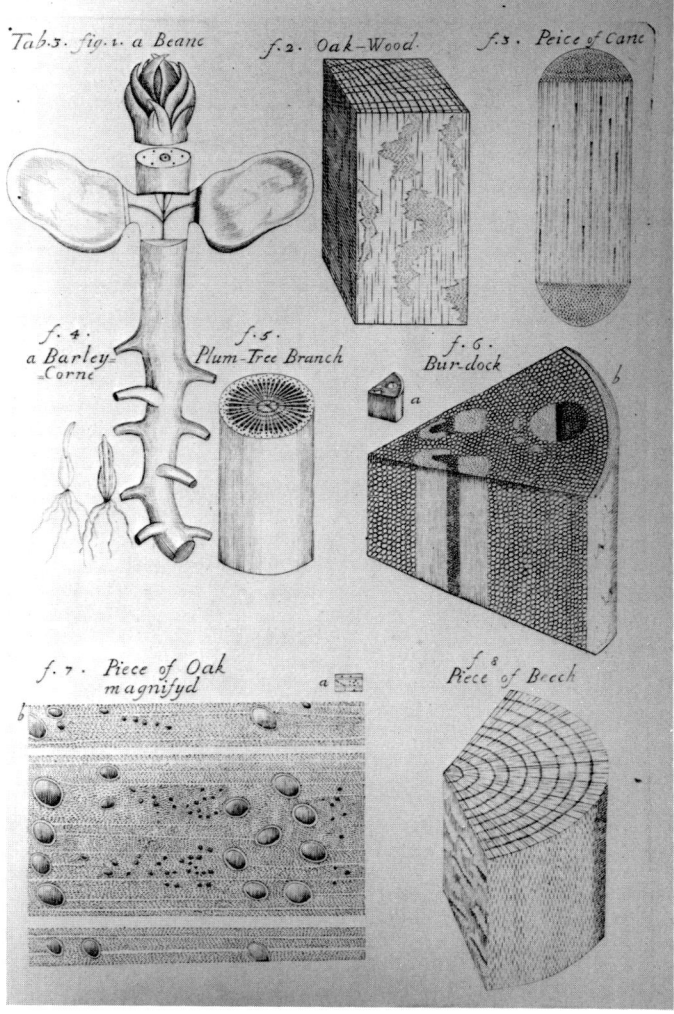

Nehemiah Grew, table III from *The Anatomy of Plants* (London, 1682). Photo courtesy of the Muséum national d'histoire naturelle, Paris.

plant structure, which came as a late fruit. This may account for the disparity between the study of human and animal life, which began in Antiquity, and the study of plant life, which began only at the end of the seventeenth century. Examination of the human body and dissection of animals yielded useful medical knowledge. By contrast, as Rousseau observed, when one is "used to looking at plants only as drugs or remedies . . . one does not imagine that the structure of the plant is worthy of attention in itself."[4] Thus, it might be said that the study of plants is a product of curiosity, provided curiosity is defined as a weakness that impels men to study what they are not supposed to want to know.

At the end of the seventeenth century and throughout the eighteenth, theoretical and practical concerns were sharply distinguished. A distinction must therefore be made between writings on plant nutrition, reproduction, and movement, on the one hand, and writings on agriculture and forestry on the other. To give one example, the treatise *De l'Exploitation des bois* (*On Using Woods*) is not to be viewed as on a par with *Le Physique des arbres* (*The Physics of Trees*). The latter treats the structure of vegetation and the mechanisms of vegetable life, whereas the author of the former has this to say about his approach: "We shall explain the point of view from which we are proposing to consider wood. Neglecting its structure, we consider it as a solid body of a certain strength and durability but susceptible to alteration and decay."[5] This difference in the focus of interest does not, of course, imply that the boundaries between the various disciplines were rigid. Between plant physiology and other subjects such as medicine and physiology relations did develop, and these relations will be described more precisely in what follows. Mere desire to understand the nature of plants in themselves was not enough, however, to allow research into "vegetable economy" to begin at once.

What new space was this, in which plants were allowed to take their place alongside animals, so near, in fact, that there was virtually no reluctance to decipher their structure? In order for the concepts and methods of animal studies to be used in the study of plants, plant life first had to be given the same theoretical status as animal life. In order to analyze plants—that is, to elucidate the parts of which they were composed—mechanistic assumptions were necessary. Now,

the seventeenth century saw a shift in the center of gravity of the universe. Its world was one in which stars and rocks were subject to the laws of mechanics expressed by means of the calculus. In order to assign a place to living beings and explain their functioning, therefore, there was but one alternative: either living things were machines in which only shapes, sizes, and motions had to be taken into consideration, or else they were not subject to the laws of mechanics, in which case it would be necessary to give up the idea that there was any unity or consistency in the world. Faced with this choice, physicists and even doctors did not hesitate for a moment: all nature was a machine, as machines were nature.[6]

This leveling of animate and inanimate objects, beings and things, had a corollary: There is a lesson to be learned from every thing in the world. Since God is the author of the world, moreover, all nature's products bear his mark. Theology wasted no time opening up new vistas. Mechanism made plant anatomy possible, but there remained a desire to provide a rational apologetics, which explains why physicists began to take an interest in minerals, insects, and plant life. "For considering that both [animals and plants] came at first out of the same *Hand*; and were therefore *Contrivances* of the same *Wisdom*: I thence fully assured my self, that it could not be a vain Design; to seek it in both."[7] Once plants became an object of interest, they gained in complexity and even dignity. "The body of a plant is just as organic as an animate body, as is evident from the anatomy that has been done on them. This fact was unknown to Antiquity."[8]

Unknown to Antiquity, yes, but unknown also to the Renaissance, when knowledge of plants was scanty indeed. When Cesalpinus and Cardanus spoke of the structure of vegetables, they merely repeated what Theophrastus and Pliny had said before them. Roots and stems, they held, were made up of parts such as bark, wood, and pith. These same parts existed in the fruit. Cesalpinus believed that the pericarp grew out of the bark, the ligneous coccus out of the wood, and the seed out of the pith. Analogies abounded: "To return to the parts of plants, all of them correspond to the parts of animals. The roots, Theophrastus believed, correspond to the stomach; I myself should prefer to see them as resembling the mouth, while the lower part of the trunk resembles the stomach; the leaves, the hair; the bark, the hide and skin; the wood, the bones; the veins, the veins; the nerves, the nerves; and the matrix, some of the entrails."[9] The seed, for its part, was described as being analogous to semen in the animal, as far as its production was concerned. The creation of a plant required the union of matter and form and presupposed direct intervention by the forces that govern the world. The liaison was effected by two intermediary agencies, the soul and the innate heat, which are located within the collar, the junction between root and stem. The heat takes hold of the nutrient in its purest and moistest state and acts in concert with the vital and vegetative faculties in order to transform and digest it. Finally, the work is completed by the specific faculty, "which prepares, fabricates, ripens, or separates the seeds and imprints upon them the resemblance, figure, and virtue of their species."[10] The absence of seeds in lower plants such as ferns and mosses is no impediment to generation. To grow them, what more needs to be done than to expose a little soil and water to plenty of heat? "Mushrooms, too, are nothing other than a kind of excrement of the soil, made to grow by excess moisture that the sun heats gently with its rays."[11]

During the second half of the seventeenth century the inner structure of the plant substance was analyzed. Only the relations between the various parts of the plant were considered, however. The approach was one of ordering empirical observations. Microscopes were of no use: "What we have performed thus far lieth, for the most part, open to the use and improvement of all men. I have described what I have observed as clearly as I can, and everything that I report can be seen without a microscope."[12] By analogy, first of all, with the distinction made in animals between the homeomeric and anhomeomeric parts, Grew distinguished between the "similar" parts (fibers, parenchyma, skin) and the "dissimilar" or "organic" parts (root, stem, leaves). Similar parts such as wood, pith, and insertions could also be organic. The next step was to study the same plant at different stages of growth. This procedure was frequently used in the study of animals, particularly insects: "That they confine not their Enquiries to one time of year; but to make them in several Seasons, wherein the Parts of a *Vegetable* may be seen in their several estates."[13] Finally, the lessons of comparative anatomy are not to be neglected: "Neglect not the comparative anatomy; for as some things are better seen in one estate, so in one *Vegetable*, than another."[14] Though plants come in a wide variety of sizes and colors, still the same structure is to be found in every species.

In regard to generation, attention at this point was focused on the species rather than on the individual plant. Plant physiologists refused to allow any such thing as "formative faculties" to play a role in generation, but the laws of mechanics were incapable of explaining how living things are formed; hence it was natural to turn for an explanation to preformationist theories. Thus generation was viewed merely as the development of a preexisting individual, which required only that the germ be activated by nutrition. For those who studied animals, the questions to be asked were obvious: Which of the two seeds contains the germ, the male or the female? Is it the female that provides the nourishment necessary to the development of the animalcule? Or is it the male that supplies food to the germ found in the egg? The same questions were asked of plants. For those who believed in the animalcule theory, the plant represented the male and the earth the female. "Plants are all in their own way like males, by virtue of their seeds and of the fact that one seed is equivalent to another, while the earth is like the female shared by all these males."[15] The ovist Swammerdam reversed the roles: "Production takes place in insects in the same way as in plants; for the seeds of plants, deposited in the bosom of the earth, incorporate themselves with the moisture in the ground; and similarly, generation is accomplished in insects by the union of imperceptible, fecund particles of the seminal fluid of the male with the seed of the female."[16] In addition, Malpighi carried out experiments, similar to those done eailier by Redi, that overturned the dogma of spontaneous generation in the plant kingdom: "I am also quite ready to believe that every plant that yields seed . . . was itself produced by seed. The great naturalist Malpighi proved this in an experiment conducted with earth taken from a deep hole and placed in a large glass vessel, the top of which he covered with a piece of silk folded several times over and stretched tight so that water and air could enter the vessel while not allowing the wind to carry in or force through any seed at all; this earth produced no plant."[17]

Was it enough, however, to set the various parts of the plant in motion for plant physiology to emerge as a discipline? No, is the only possible answer, for a very simple reason. The procedure by which the various parts of the plant were distinguished could not be extended in any obvious way to an attribution of particular uses to those parts.

It was easy enough to study the wood, the insertions, the fibers, or the leaves, but nothing in their form made it possible to deduce their function. By contrast, physiological observation of organ function is inseparable from the study of organ structure. Some writers therefore began to suspect the existence of another anatomy.

The following point has not, it seems, been sufficiently stressed in the literature: There is one physiology of plants but two anatomies. One plant anatomy is merely an orderly exposition of empirical observations of plants, nothing more; the other, inextricably bound up with physiology, is a copy of animal anatomy. To be more precise, the word *physiology*, as a term used to denote the study of plant functions, appears for the first time in the work of Linnaeus. The author of the *Botanic Philosophy* distinguishes between two kinds of work by botanophiles, research in anatomy and research in physiology: "In anatomy: Malpighi, Grew. In physiology: Feldmann" (Aphorism 44). Linnaeus' proposed distinction is revealing, especially when we look at the titles of the works published by these authors. Grew: *The Anatomy of Vegetables, Begun, with a General Account of Vegetation Founded Thereon* (1672); Malpighi: *Idea anatomes plantarum* (1672); Feldmann: *Dissertatio physico-medica inauguralis sistens comparationem plantarum et animalium* (1732).

Three observations are called for. First, if plant physiology is based on comparisons between plants and animals, it is clear that there is no continuity between this physiology and the anatomy of Grew or Malpighi. For, as we have seen, this anatomy was merely an orderly exposition of empirical observations of plant life.

Second, since organs are defined by both their structure and their function, there must be another anatomy, quite distinct from that of Grew and Malpighi, and in a sense in parallel with it. What anatomists like Grew and

Malpighi called fibers and leaves, for example, this second anatomy would call veins and lungs.

Finally, although the term *physiology* appeared only in the middle of the eighteenth century and received a definition only toward the end of that century ("I have preferred this word to indicate the science that reveals to us the organization of plants and the history of their life, rather than the term 'vegetal physics' which M. Duhamel adopted, because the meaning of the former seems to me more precise than the meaning of the latter"[18]), plant functions had already been treated not only in Feldmann's dissertation but also in the writings of Grew and Malpighi. Since plant physiology was not dependent on an analysis of plant structure, my next order of business must be to show what it was that made it possible to talk about how the various parts of plants are used.

THE "VEGETAL" MODEL

That there is no logical relation between Grew's attempt to give an orderly exposition of empirical knowledge of plants and his work in physiology should not be surprising. When interest revolves around the fibrous structure of the plant, the plant's economy will inevitably be conceived in an ambiguous way. On the one hand, the plant is seen as a pleasant bit of architecture: "How admirable also is the natural Structure or Organism of Bodies? The whole Body of a Plant, whether Herb, Shrup, or Tree, is composed of two Species of Fibers, so artificially managed; that all the Parts, from the Root to the Seed, are distinguished from one another only by the different Position, Proportion, and other Relations and Properties, of those two sorts of Fibers."[19] On the other hand, it is also recognized that the construction of the plant, admirable though it may be, raises certain problems. In regard to structures and functions, the plant kingdom lags well behind the animal kingdom. However artful the assemblage of fibers may be, as a model of living beings it is all too rudimentary. The study of plants differs from animal physiology in that no part of the plant calls up the idea of an organ, a function. What attitude is the observer to take? Should he be satisfied with his first impression? In that event plant studies would be over almost before they begin, for the plant would be an organism without organs. Or, alternatively, should the investigator suspect that plants harbor within themselves organs as differentiated as those found in animals? Faced with these alternatives, all physiologists chose the second course, inasmuch as growth, reproduction, and motion are phenomena common to both plant and animal life, and plant structure must accordingly be more complex than it appears at first sight. The study of plants is a delicate task, because the problem is to apprehend the plant organs and understand their functions. A fixed point of reference is necessary, for an ill-defined and obscure domain can be explored only with the aid of the light shed by knowledge of a relatively well-known and clearly defined subject area. Thus animal life must serve as the model for plant life.

Someone might well object that the idea of understanding what is known less well in terms of what is known better is not a new one. In Antiquity, Aristotle used this very method to determine the parts of plants and their uses. Insufficient stress has been laid, however, on the disparity between his intuition that the plant kingdom belongs to the world of living things and the very rudimentary structure that he attributed to plants. This disparity was apparently the result of a compromise. For it is clearly impossible to reconcile the contradictory requirements of trying to explain the plant kingdom in terms of the animal while respecting the hierarchy of functional principles. Nutritive and generative principles are appropriate to both plants and animals it is true, but the sensitive and locomotive principles guide the actions of animals alone. Nutrition in the animal therefore involves the existence of a sense of hunger, a faculty of choice. Generation, for its part, requires a locomotive function in distinct individuals; sexual union presupposes separation. Thus it is scarcely possible to conceive of organs and functions in plants dependent on the effects of the sensitive and locomotive principles. This explains why Aristotle's discussion of the nature of the plant kingdom is more or less completely contained in its defintion: "It is chiefly man who, owing to his upright posture, has the privilege of situating himself in the same sense as the entire world. As for plants, which are immobile and which draw their food from the soil, this part must be down below, because the roots are analogous to the part that is called the mouth in animals. It is through the mouth or root that the one receives food from the soil, the other on its own."[20] To say that the plant is a man turned upside down is to claim that the plant possesses a "nutritive soul." But the faculty of sensitivity is lacking in plants, so that it is more correct to compare the roots to the lacteal ducts. "For plants get their food from the earth by means of their roots; and this food is already elaborated when taken in, which is the reason that plants produce no excrement, the earth and its heat serving them in the stead of a stomach."[21] The production of fruit, moreover, is a consequence of receiving nourishment, since plants remain fixed in one place and therefore have no sexual life. This explains why "the nature of animals that do not move . . . resembles that of plants, [having] no sex any more than plants have."[22]

Nothing of all this changed during the Renaissance, which was satisfied merely to rehash the ancient doctrines. When Cesalpinus discusses plant nutrition, he repeats what the ancients had said before him. His problem was essentially to reconcile the views of Aristotle and Galen. One could agree with Aristotle that the root is the superior part and therefore that the plant is an inverted man. But one could equally well agree with Galen that the plant was an animal turned inside out: "Just as the nutritive principles mount within the bodies of animals, so, too, do they mount within vegetables. . . . It therefore becomes necessary to place the root at the bottom of the plant, the stem above; for, in animals, the veinous network also begins at the bottom of the stomach, and the principal vein rises from there to the heart and head."[23] In regard to generation, it was common during this period to refer to the manner in which palms are fertilized to prove that in the plant kingdom the sexes are generally distinct from each other. A mixture of travelers' tales, ancient texts such as those of Herodotus and Pliny, and fables were used for the purpose. Duret in fact held that "the dust" was not the only agent of fertilization. The male impregnates his spouse with "his breath and glance." Fertilization also takes place via the cord that binds one tree to the other; the female, "who thereby senses heaven knows what secret communication from him to her—a message that flows insensibly from the one to the other (just as the numb fish transmits its venom along the rod that touches it, putting the hand and arm of the fisherman to sleep)—the female takes pleasure from his message and lifts up her branches."[24] Many other plants are also united by such relationships and mysterious affinities. In everything the inclinations, tastes, and appetites of the partners complement each other. The grape vine loves young elms and poplars, for, "having married these trees, the vine spreads its tender shoots and delicately climbs, in this way taking its pleasure and bringing forth bounteous fruits."[25] Love between palms, then, is not unusual; plants are involved in relations similar to the relations between animals. Often, moreover, the distinction between male and female is merely metaphoric. One peony is called male because it is used in concocting a remedy intended for use by men, while another is called female because it is to be used exclusively by women. The appearance of a plant, too, sometimes exhibits a distinctive character. "The first species of so-called male hemp has a round stem and grows to some four or five feet in height, branches a great deal, and resembles a shrub"; the female looks similar, "but the stem is thinner and simpler, with no collateral branches."[26] In some cases the criterion of femaleness was the beauty of the flower: "The cluster of flowers [of female balsamita] is lesser than that of [male balsamita], but of a better smell, and yellower colour."[27] Clearly, during the Renaissance no one had any specific idea about how plants are fertilized.

The method used in the late seventeenth century and throughout the eighteenth century was the same, however. That is, the significance of the unknown was interpreted in light of the known. Why did this same method now turn out to be fruitful? Why was the recognition that plants are living things now matched by an understanding that their structure is as complex as that of animals? It is because the status of nutrition and generation underwent a change in the classical era of botany, and vital phenomena were subsumed under mechanism. Descartes had earlier argued for a conception of the animal in which no part was played by the notion of a vegetative or sensitive soul. La Mettrie said almost the same thing: "By the vegetative soul the ancients meant the cause of generation, nutrition, and growth of all living things. The moderns, who have paid little heed to the idea these early masters had of this species of soul, have confused it with the very structure of animals and vegetables."[28] The central issue shifted from the vegetative soul to vegetation. In order words, the issue became one of a mechanism rather than a soul, a faculty, or a principle. The word vegetation referred to the same thing in animals and plants, namely, the mechanism responsible for nutrition, growth, and generation. This is a crucial point. How do we distinguish between interior and exterior in animals, between the entrails and their envelope, if not by viewing all the organs together as constituting the vegetative apparatus? Even more, how else do we distinguish between the vegetative life of animals, and the animal life as such, which is manifested in the actions of the senses and locomotion? "Vegetable life is independent of sensible life." The consequences of this cleavage between "vegetable life" and "animal life" for the study of plants are immediately apparent. Not only does it become possible to stake out common ground in studying both classes of structured beings, but, too, analogies can now be established between plants and animals, because "it is the same Vegetable Life, whereby these Operations [Nutrition, Augmentation, Generation] are performed in Animals, as well as in Plants."[29]

If the possibility of establishing a correspondence between vegetable economy and animal economy exists, however, it still remains to be shown that such an approach is necessary. Why should the animal kingdom be chosen at the outset as the term of comparison, when the knowledge sought is of vegetable structure? For one thing, practical experience with technology often "communicates some of its structures to the perception of organic forms,"[30] particularly animal forms. This, at any rate, is the answer that emerges from a study of the vocabulary of animal anatomy in Western science. Neither the inner nor the outer structure of the plant lends itself to comparison with the artifacts of technology though. "Plants are distinguished from nature's other productions by the singularity of their structure, among other things."[31] Another reason for choosing animals as a term of comparison has to do with the nature of the plant as an object of study, given the existing state of experimental techniques. In animals the internal structure is easy to unravel, and the various organs are clearly distinguished. Their form and color vary, and it is easy to see how the various parts are arranged and connected. By contrast, the structure of plants is far less differentiated, and the parts are not easy to identify. Dissection is of little use. "The vegetable anatomist can resort to dissection, maceration, dissolution, and natural injections, that is, the art of making vegetating plants imbibe colored liquids. These means are not, however, as effective in plants as we know them to be in animals."[32] Dissection soon meets its limits, owing to the crudeness of the instruments; maceration alters the parts of the plant; and injections are not correctly calibrated to the diameter of the vessels. Nor can the speed of fluid circulation be increased by force, because uprooting the plant kills it. In short, then, metaphor proves an inadequate basis for discussion of the parts of vegetables, and the methods of anatomy are no help.

How can one hope to get beyond the apparent simplicity of the plant, which impedes observation, unless familiar mechanisms can be found at work in the plant kingdom? How can the enigma of the plant be penetrated without the help of anatomy? There is only one way out: an approach opposite to that of animal physiology must be tried. Structures must be deduced from the study of functions. Animal physiology had been based on deduction from the anatomical structure of animals, defining itself, in Haller's phrase, as "anatomy in motion." Plant anatomy, by contrast, should be based on deduction from physiological function and should define itself as frozen physiology. The reason for this reversal is immediately apparent. The purpose of the various parts of the plant are not at all clear; hence there is no choice but to focus on the phenomena of growth, self-preservation, and propagation, which demonstrate that the physiological functions are in fact being carried out by the plant. Anyone who looks at plants in this light must look to zoology, where models abound. This explains why animal physiology enjoyed epistemological priority over plant physiology. In the classical era, investigation of the plant revolved around knowledge of the animal functions.

Physiologists who were looking for functions in plants similar to those observed in animals immediately ran into a problem. For, if there were too great a similarity between the two classes of structured beings, the kinship between them might be destroyed. The plant kingdom, in other words, might cease to be seen as a distinct object of study. Further, it was assumed that there was a hierarchy of modes of organization in living things; if plants and animals were too nearly alike, the ranking determined by this hierarchy would no longer be respected. If, instead, one assumed not that plants were similar to animals, but rather that nature followed different models in constructing the two kingdoms, the result would be the same. Unless one could discover what the model for the plant kingdom was, it would be impossible to study plant functions. Accordingly anyone who wished to discuss the nature of plants had to assume that they were neither totally different from, nor faithful copies of, animals. One could meet both these requirements by basing the study of plant physiology on two complementary principles. The first was necessary if plant physiology was to avoid merely duplicating animal physiology: "Nature is too complicated in its simplicity, too varied in its uniformity, to permit an observer to have full confidence in the general ideas to which it gives rise."[33] The second gave botanists the right to base their arguments on similarity: "*Like effects must have like causes.* This rule is based on the uniformity of Nature's processes; if a cause is sufficient in one case, why should it not be sufficient in another, perfectly similar case?"[34]

Plants, then, could be seen to be different enough from animals to allow a different anatomical substrate to underlie similar functions. At the same time, similarities between the two kingdoms were evident. This explains why analogy was accorded so important a role: "How should an observer proceed with his observations, if the subject is completely new? ... His steps should follow the path marked out by analogy between his subject and others about which something is already known."[35]

To be sure, the work of physiologists would be whimsical indeed if they were to content themselves with establishing analogies. The main point is to state the terms of a primary problem. "In order to give the analogy the utmost probability, the properties in respect of which the effects or causes believed to be analogous resemble one another must be established by observation."[36] In this period, *observation* meant merely familiarity through the senses. Physiologists gave first priority, of course, to the sense of sight. Since plants are composed of solid and liquid parts, importance was attached to the study of shapes, the arrangement of parts, and the color of fluids. Touch and taste were also used, because the flavor of liquids and texture of solids were characteristics that might furnish useful indications as to the nature of the object under study. Sénebier went so far as to contend that there is an "observational spirit," a "faculty of carefully perceiving each object as it is in itself, grasping all its relations with other beings, linking them all together, and giving a faithful portrait of the whole."[37] To sum up the procedure, the naturalist first establishes an analogy (chooses a model). Next, he selects a method of observation (a marking technique) that will enable him to give a full description of his object. No matter what model and technique he chooses, however, the naturalist completes his work by presenting a portrait that is supposed to show what the object of observation is.

If observation alone were enough to lay bare the true nature of plants, not only would analogies be superfluous, but in the final analysis plant physiology would not exist. For ordinary language would be the natural vehicle of knowledge. In reality, observation leads the naturalist to construct the economy of the plant in such a way that gaps remain to be filled and problems remain to be solved. Gaps remain because the observer will inevitably find some areas rather obscure; thus "observation leads to experiment by stimulating curiosity; it makes one search for ways to fill in the blank spots." Further, observation raises problems, because there is no way to be sure that the characteristics one sets out to uncover are reliable indicators of the nature of the plant parts. Again, it is "necessary to combine experiment with observation, because nature cannot be interpreted in too many ways. Experiment is certainly one of its keys. Experimentation brings out many particular relations between beings and identifies causes and effects, gauging them precisely whenever possible."[38]

One form of vegetable anatomy emerged at the same time as the Cartesian system. Grew and Malpighi unraveled the structure of vegetable matter merely by imposing a certain order on its component parts. Before this, however, undergirding this anatomy as a kind of counterpoint was a form of physiology in which a different version of anatomy was implicit, an anatomy of plants that was really a copy of animal anatomy. The existence of these two anatomies no doubt accounts for the problems that Julius von Sachs encountered in writing the brief eighth chapter of his book, entitled "Botanical Anatomy in the Eighteenth Century": "In Italy, Malpighi had no successors worthy of the name; in England, the fleeting splendor the science had enjoyed while Hooke and Grew were on the scene soon vanished without a trace, and it might be said that the ensuing lackluster period has persisted down to the present day."[39]

According to the mechanistic view, it is not faculties that are responsible for the preservation and propagation of the species, but rather functions. Thus there is nothing to prevent one from establishing analogies between plants and animals. Animal physiology is prior to and serves as a model for plant physiology, because the internal structure of plants is hard to perceive: the mechanism of vegetation is far more secret and obscure than that of animals. In short the approach taken to plant physiology was the reverse of the approach to taxonomy, where plant life enjoyed priority. Since the external parts of plants are visible and easily identifiable, classificatory botany is prior to and serves as a model for zoology.

If the mechanism of plants can be grasped only to the extent that it reflects what is already known about the mechanism of animals, however, it follows that the animal world is to plant physiology as the technological world is to animal physiology. Ultimately this comes to the same thing as saying that the technological world, via animal physiology, structures the perception of plant phenomena. The attempt to explain the inferior in terms of the superior is merely a form of mechanism, not an indication that mechanism has been dismissed.

2 Nutrition

By the late seventeenth century, studies of nutrition in plants and animals had revealed a disparity. In animals the nutritive function was associated with a series of well-defined organs. By contrast, the apparent simplicity of plants was disconcerting. Research to find out how nutrition is accomplished in plants was thus imperative. The problem was not only to discover what mechanisms are involved in absorbing, transporting, and preparing the food but also to study how growth takes place.

To inquire whether these mechanisms exist and, if so, to ascertain their nature, models based on the animal functions had first to be selected so that the presence of essential functions in plants could be established by analogy. Choosing a model was at once easy and difficult—easy because animal physiology abounded with possible candidates; difficult because the choice of models that could be transferred from the domain of animal physiology to the domain of plant physiology was presumably not arbitrary. The way to overcome the difficulty was to mark out a certain domain within animal physiology, which came to be known as "vegetation." This necessitated a choice. Either one could hold that vegetation involved the action of a definite and limited set of mechanisms, in which case the thing to do was to look for organs in plants that might serve as seats for these same functions; or else one could define vegetation more broadly, as a continuous cycle of assimilation and disassimilation, in which case it might be possible that the structure of plants is different from that of animals. Those who chose the former course created what I shall call, to use Jaucourt's phrase, *plant mechanism*; while those who chose the latter worked out what I shall call the *system of vegetation*.[1]

In histories of botany, a contrast is drawn between the mechanists and the vegetationists similar to the one drawn between the proponents of analogy and the advocates of observation and experimentation. The mechanists identified the roots with the lacteals and the leaves with the lungs and thus found what they considered to be circulatory and respiratory functions in plants. The vegetationists, on the other hand, were physicists who applied the scientific method to the study of plants. Boerhaave, Malpighi, and Perrault went astray, so we are told, because they looked for analogies. Not only did they fail, according to the historians, to perceive the resemblance between plants and animals with respect to the important function of nutrition, but, as mechanists, they were presumably incapable of proceeding in any other way. In other words, their adherence to mechanism precluded their looking upon plants as living things. By contrast, the use of an experimental method and the willingness to submit hypotheses to experimental verification are supposed to have enabled Hales and Guettard to state the first laws of vegetation. We are told that these two men demonstrated that plants and animals nourish themselves in similar ways. Finally, as vitalists, they are supposed to have seen that certain aspects of vegetation are irreducible to physical and chemical laws. The heart of the question, however, lies not in this series of contrasts but in the possibility of constructing two alternative versions of the nutritive function.

For mechanists, circulation was apparently a crucial function in plant nutrition. Harvey published *De Motu cordis* in 1628. Thus some forty years separate the discovery of circulation from its use as a model by Major, Mariotte, and Perrault. Perfectly good reasons existed for this delay. Although Aselli had discovered the chyliferous vessels in 1625, it was not until 1647 that Pecquet observed and described the thoracic duct that empties digestive products into the left subclavian artery. Hence the chyliferous vessels did not terminate in the liver, as Aselli had thought. This organ was therefore stripped of its ancient privilege. What organ was responsible for performing the important function of fabricating the sanguinary fluid out of the substances flowing in from the lacteals, if it was no longer the liver? Primary importance was now given to the heart. It was the heart that distributed blood to all parts of the body. More than that, though, it was also the heart that was said to "elaborate" the nutritive matter: venous blood was raw nutriment that had to circulate before it could be sufficiently digested and worked into the proper consistency to provide nourishment. Reasoning by analogy, it was possible to transfer this physiological model from animals to plants. The sustenance and growth of animals was based on the preparation of the nutrient through circulation. Plants, too, sustain themselves and grow. Therefore, "the vegetation of plants, that is, the appropriation of the nutritive juices, can ordinarily be carried out in the plants themselves, in consequence of which their life cannot be sustained without provision of balsam and without transit of the superfluous matter. Thus, within plants themselves, there is undoubtedly some sort of circulation."[2]

When vegetables are examined, however, no organs similar to those disclosed by anatomy in animals are immediately apparent. All we need do to overcome this difficulty, however, is to examine the properties of plant life for signs of the nature of the various parts of plants. A priori, it may be said that these properties are of two kinds: consistency, in regard to containers, and color, in regard to their contents. In animals arteries are firm and rigid, but veins are soft. As for red blood, the nutritive humor, its cast is more vivid than that of venous blood. If one follows Perrault in adopting an identification technique based on the consistency of the container or receptacle, then the vessels in the wood must be seen as comparable to the arteries, and the vessels in the cortex as comparable to the veins. Major made a similar choice: "The roots of plants, the wood, the cortex, the leaves, the flowers, and the fruits contain fibers, some of which correspond to the veins and others to the arteries in the body of an animal, depending on their specific degree of hardness."[3] The parts will be located in a diametrically opposite way, however, if one follows Mariotte in choosing an identification technique based on the color of the liquids: "Not all the humor contained in these plants was colored, but only that part which was contained in certain vessels that I should compare to arteries. . . . That which is found in the remainder of the stem is related to the blood contained in the veins."[4] The use of these methods also brought other benefits. It became possible to specify the direction in which the sap circulated and to indicate the part of the plant responsible for cooking the humors, a function performed in animals by the heart. For Perrault, who associated qualities with substances, the undigested, heavy sap follows the path of the soft cortex and flows toward the roots. By contrast, the hardness of the wood corresponds to the volatility and hence purity of the nutritive substance, which has a natural tendency to rise. "It is assumed that this takes place in the same fashion as in animals, where the arterial blood leaving the heart, which is to the animal what the noblest part of the root is to the plant, is distributed throughout the body, which, having retained that portion of the blood appropriate to its sustenance, returns the rest to the heart."[5] Conversely, for Mariotte, the nutritive humor analogous to the red blood originates in the leaves and descends through the cortex, and the undigested blood returns to the foliage through the vessels in the wood. Clearly, the differences between the systems put forward by Perrault and Mariotte can be accounted for by their use of such crude methods.

If such methods are to be useful at all, there must of course be a marked difference between wood and cortex. In many plants, however, there is not even the slightest feature to allow distinguishing between veins and arteries. This poses a problem for the botanist who wishes to argue that the sap circulates in all plants. The analogy with the animal kingdom can always be modified, though, so that the term of comparison becomes the lower rather than the higher animals. "Just as there are some animals in which the vessels are clearly distinct, while in other, less perfect animals (including most insects), not only veins and arteries but also heart and liver are invisible, so we may say that there are plants in which the circulatory organs are distinct and visible . . . and others in which the vessels and pathways are hidden and unknown."[6] Dedu borrowed this argument from Perrault, who went so far as to imagine circulation without organs: this involved the commingling of two humors, one nutritive, the other undigested. Thus nothing precluded the belief that circulation is common to all living things, plants and animals alike. This being the case, it would have been erroneous to conclude that circulation did not exist merely because the senses did not perceive it.

Whether or not it is possible to identify different kinds of vessels in plants, to follow their paths and trace their interconnections, a further problem remains to be resolved. In the higher animals, the food is distributed by a propulsive force not found in plants. As far as we can see, no part of the plant contracts in an obvious and forceful manner, as does the heart. In order to understand the motion of the sap, then, we must imagine how nature might remedy this defect. Only two answers are possible: either the "principle of motion" lies inside the plant, or it lies outside it. Mariotte chose the former solution. The vessels, he argued, must exert an attraction on the fluids they contain, and the phenomenon of imbibition must also play a part. Perrault took the second option: his explanation involved the action of the wind on the branches, whose swinging back and forth compresses the vessels and hence also the fluids they contain. Of course this compression was quite incapable of determining the direction of flow, at least in the case of the humor analogous to venous blood. Perrault was therefore forced to turn to the theory of circulation for help. To ensure that the undigested sap would flow toward the root, the vessels of the cortex, being the analog of the veins, must be equipped with parts that act as "valvules." Mariotte, on the other hand, believed that the nutritive sap descends via the cortex. Hence he differed with Perrault and situated the valvules in the wood rather than the cortex: "These conduits are apparently arranged in such a way that whatever enters into them is prevented from flowing out; we find similar examples in several other parts of the body, and even in the veins, where there are small pieces of taut skin, known as valvules, arranged so as to allow the blood going to the heart to pass but preventing flow in the opposite direction."[7]

The circulation of the sap, conceived by analogy with the model of circulation in animals, was put forward as an obvious fact rather than a hypothesis. It can be seen, a priori, that the evidence adduced could not but confirm the theory. In transferring the proofs for circulation of the blood to the question of circulation of the sap, however, the circulationists had to make certain adjustments. In the first place, it was possible to reverse some of Harvey's arguments. Once circulation had been discovered, for example, it became possible to explain how substances applied to a body at one point could propagate their effects, whether for good or for ill, throughout the organism. This, however, is a proof of circulation based on its consequences, an argument a posteriori. Assuming that like effects stem from like causes, it follows that "the corruption that migrates from a damaged part throughout the entire plant cannot be explained without circulation."[8] But this is an argument a priori. Or again, to take another example in which the argument was turned around, Harvey had referred to the similarity between the microcosm and the macrocosm in discussing circulation; he employed this argument, however, by way of confirmation through analogy, and only after he had discovered the circular flow. Perrault, on the other hand, described the circulation of the sap as a special case of circular motion, which he looked upon from the first as a universal model. Plants cannot do without circulation, "since it is necessary to the sustenance of still less perfect beings, which are not plants. The juice contained in the earth is doubtless less perfect than the plants that are nourished by it; yet this juice could not possess its peculiar perfection if it did not circulate constantly: for it must rise into the air in the form of a vapor, there to be cooked as much by the heat of the sun as by the agitation of the winds, which first separate and then mingle together its different parts, after which it falls back down to earth."[9]

The facts could also be interpreted so as to fit the theory: if, like Major, one followed Harvey's line of argument, the procedure was first to identify the earth with the heart. Circulation, for Harvey, had answered the question, What becomes of the large quantity of blood that flows forth from the heart as from a spring? Applied to the plant kingdom, the theory of circulation provided an answer to the same question:

If it were merely a question of motion of the kind spoken of earlier in the bodies of plants—just as some people used to hold the overly simple belief that in the animal body the old blood flows toward the outer parts via the arteries and veins—by such means vegetables . . . would be unable to feed themselves readily or to grow, because more than the appropriate amount of juices would constantly be flowing in . . . to the point where the daily influx of nutritive substance would overwhelm the plant. But since vegetables endure for a certain period of time and mature without difficulty, there can be no doubt that the nutritive juices describe some sort of circular trajectory within them, by means of which the appropriate nourishment is carried to each part of the plant through specific vessels, while the excess is evacuated by other vessels.[10]

The next step in the argument was to identify the vessels in the cortex with the veins, if, like Perrault, one was seeking to confirm the existence of circulation by experiment, the results of which were interpreted in such a way as to reaffirm what had already been incorporated into the structure of the plant through observation. The turgescence of the cortex just above a ligature shows that this part of the plant "conveys the juice flowing back to the root, and that the rising juice, which flows more deeply within the plant through hard and fibrous vessels, is not impeded by the ligature, which compresses only the outer portion; this is analogous to what happens in bloodletting, where the ligature stops the flow of blood in the veins only, owing to the weakness of their covering."[11] In short, these summary comparisons had a quite definite function, to obscure the differences between plants and animals so as to create the illusion that Harvey's approach was being followed.

The more closely anatomists examined animals, however, the more complex their organization appeared to be. The idea that the preparation of the nutriment is accomplished solely by circulation became untenable. For one thing, the work of Wharton and Glisson revived the old distinction between spermatic and sanguine parts, modifying it in one respect. The white parts, it was argued, are nourished not by a prolific humor, sperm (which was supposed to have priority over the blood), but rather by a humor prepared by the nerves and carried by the lymphatic ducts. For another, Malpighi's research on the structure of the lungs led him to argue that the lungs are responsible for "sanguification," the final phase in the elaboration of the nutriment. Thus animal physiology was not long in providing new models for plant theorists to choose from. If, like Grew, one chose the first theory, it followed that "some portion of the *united Principles* both of the *Parenchymous* and *Lignous Parts*, [is] necessary to the true *nutrition* of each: As the confusion and joynt assistance of both the *Arterious* and *Nervous* Fluids, is to the nourishment or coagulation of the *Parts* in *Animals*."[12] The problem, then, was to identify the organs of nutrition in the plant. But this turned out to be difficult to do. Whereas in animals it was easy to distinguish between lymph ducts, blood vessels, and nerves, in plants the different parts were intimately associated, not to say inextricably intertwined. Grew could do no more than his predecessors to overcome this problem and lay bare the network of vessels of different kinds. If he achieved results apparently more precise and coherent than they did, the reason is that he looked at the receptacles and their contents in a somewhat different way from Perrault and Mariotte. Instead of studying the consistency of the parts, Grew studied their structure. And instead of looking only at the color of the fluids, he gave a fuller analysis of their physical properties. He did this because in animals the blood vessels are of larger diameter than the lymph ducts. The former, moreover, contain a nutritive humor, while the latter contain lymph, which is clear, white, and tasteless. Do we not observe similar vessels in plants? Some resemble the blood vessels and others resemble the lymph ducts: "The *Lactiferous Vessels* are tubulary, as the *Lymphae-*

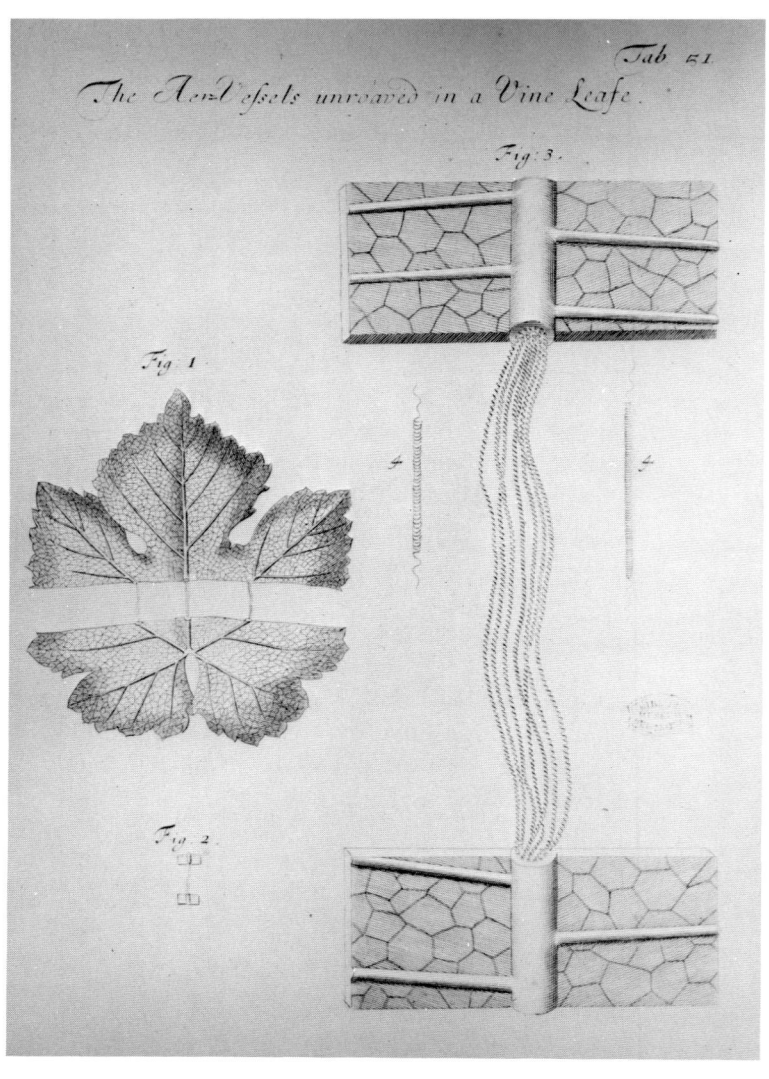

Nehemiah Grew, table LI from *The Anatomy of Plants* (London, 1682). Photo courtesy of the Muséum national d'histoire naturelle, Paris.

ducts, but of a somewhat wider *Concave* or *Bore*. . . . They are more adapted to the free motion of the *Milky Content*: which [is] an Oleous and Thicker *Liquor*, than that in the *Lymphaeducts*."[13] Like Perrault, Grew used the notion of pressure to explain the mechanisms responsible for the rising of the sap. For Grew, however, the pressure that acts on plant vessels was the result of dilation of the parenchymal cells rather than of wind-activated motion of the branches, as Perrault had maintained.

Malpighi, who had discovered secondary circulation in animals, believed that the sap must spend a certain period of time in parts of the plant analogous to the lungs. The lungs played an essential role in mixing the different humors. Of course, respiration, the movement of the chest that causes air first to be drawn into and then expelled from the lungs, depends on muscular action. In plants, however, there are no muscles. Hence the analogous parts of the plant cannot possibly have the same structure as the lungs. In particular, there is no reason to expect to find air cells capable of expanding and contracting so as to stir the humors, or heat to cook them. At this point a problem analogous to Perrault's cropped up. The absence of a heart in the plant forced Perrault to look outside the plant for the "principle" of the sap's motion; similarly, Malpighi was obliged to situate the agency responsible for stirring the humors and the source of heat responsible for cooking them outside the plant. These factors could then, of course, affect only the outer parts of the plant. Doubtless this explains why so important a role was assigned to the leaves. "In conclusion, I believe it probable that nature intended that leaves serve the following purpose, namely, to allow the nutritive juice flowing from the fibers of the wood to be cooked in the utricles . . . pulverized by the force of the sun's rays, mixed with the existing matter remaining in the utricles, allowing growth of new parts. . . . The change is the same as that undergone by the new nutrient absorbed by animals when it is diffused throughout the blood left in the vessels by earlier nutrition and transformed by it into a blood of uniform nature."[14] Malpighi based "sapification" as well as "sanguification" on a model drawn from the kitchen, the baking of bread. The "nutritive juice" and the chyle are analogous to the dough; the "existing matter" and venous blood correspond to the yeast. Sources of heat, the sun and the heart, correspond to the oven. The sun's rays and the air trapped in the pulmonary air cells assure the mixing of the humors, just as the baker's kneading works the yeast into the dough so that the bread will rise.

Of course the lungs also have other functions in the higher animals. They give rise to the motion of the fluids. To begin with, the inflated air cells exert a pressure on the blood that drives it toward the heart. Then, too, the contraction of the air cells depresses the diaphragm, thereby compressing the lacteal ducts; this forces the chyle to rise into its reservoir. The leaves are surely well placed to distribute nutrient to all parts of the plant. For the sap flows out of the utricles, just as the blood flows out of the lungs. But the foliaceous parts cannot cause the sap to flow into the roots or to ascend through the stem ducts. Thus problems arise in attempting to explain these phenomena. Study of plants can eliminate the difficulties, however. The solution involves observation of a part whose structure is similar to that of the trachea in insects. Malpighi therefore designated this part by the term *tracheae*: "These fibers are composed of a small, narrow, transparent, silver-colored, spirally wound strip. In the places where they are joined, the edges of this strip cause the fiber to be uneven inside and out. If the end is detached and the strip taken apart, it does not appear to be made up of separate small rings, like the tracheal artery of the perfect animals. Rather, one notices that it is a long band, like that which forms the trachea of insects."[15] Since the tracheae sit between the wood and the pith, Malpighi assumed that they were linked to the fibrous vessels and served to convey the sap. To explain the motion of the tracheae, it was apparently necessary to rely upon the alternation of periods of heat and cold, day and night. The thinning of the air in the tracheae causes their volume to increase. A pressure is thereby exerted on neighboring vessels, causing the sap to rise. Conversely, the condensation of the air relaxes the tracheae, leaving the vessels ready to receive fresh juices. In order to prevent the sap from flowing back down the stem, however, it was also necessary to assume that "the portions that join the fragments of fiber together project slightly inward and thereby play the role of *valvules*, so that even the smallest drop is forced to rise to the highest extremity of the plant, as though by a rope or staircase." [16]

The difference between Grew and Malpighi is now clear. Grew used the same words to denote the plant vessels as were used to name the vessels in animals. He did so because their structures seemed to him identical. The plant is composed of an assemblage of lymph ducts and vessels proper, and "what the *Viscera* are in *Animals*; the *Vessels* themselves are in *Plants*." [17] Accordingly, the nutrient stemming from the earth acquires its perfection in the plant. The *cambium*, another term borrowed from the vocabulary of animal physiology, results from the commingling of a humor analogous to the blood with a nervous or vital juice. Malpighi described a functional correspondence between the lungs and the leaves, but by contrast he said that the plant differs from the animal in that the nutrient is prepared for the plant by external organs. The nutrient that is the analog of red blood is perfected on the periphery of the plant, as the sap is mixed with a juice that has remained for a period of time in the leaves. In the final analysis, the two conceptions of plant nutrition are opposed to each other. They are opposed because they are based on divergent initial choices of model from among the theories advanced by animal physiologists.

Regardless of whether nutrition is accomplished by circulation, by the commingling of juices in the vessels of the plant, or by the cooking of those juices in the leaves, a further problem remains to be resolved. Nutrition cannot take place without absorption and excretion. Physiologists were therefore under the obligation to explain how these functions are performed in plants. The selection of models was determined by the way in which plant nutrition was conceived. The solutions proposed by the circulationists Grew and Malpighi were different but shared one point in common: the source of the nutriment that the plant has to elaborate is a substance analogous to chyle. This is no doubt the reason that the roots were identified with the lacteals. Structurally they are the same, and both take in liquids. "It is known that the substance that nourishes animals, after preparation by the stomach, passes into the bowels, where it encounters imperceptible small pores and vessels through which the most subtle parts of the substance known as chyle pass and enter. . . . Now, on the basis of this analogy between vegetation in animals and vegetation in plants, it seems likely that the rainwater, mixed with the other principles of which plants are composed, being joined to and contiguous with their roots, comes into contact with imperceptible pores, through which it enters the plant."[18] As for excretion, it is possible to imagine a cleansing of the sap similar to the cleansing undergone by the blood, that is to say, perspiration. The plant absorbs a refined food analogous to chyle, so that it does not need to eliminate large pieces of excrement, as animals do. The leaves of the plant, it is then argued, are like the skin of animals. They are subjected to the effects of the heat, which encourages excretion via perspiration; and their structure is similar to that of the skin. "I have long asked myself to what animal organ the leaves are analogous. . . . I have been largely of the opinion that they fulfill the functions of the skin or hide. . . . It is then easy to see analogies in the structure of leaves and the uses to which they are put. In the leaves, according to a certain economy, there are curious vessels, namely, the tracheae, the ligneous fistulae, and their associated ducts, which terminate at the surface; thus sweat ducts and perspiration are not lacking, and there is sometimes even hair."[19]

Throughout the eighteenth century, at least until a form of chemistry emerged as a discipline, the study of nutrition continued to revolve around questions whose outlines, largely laid down at the end of the preceding century, have often been described. Observation focused first on the respiratory and circulatory functions and second on the functions of absorption and excretion. Plant growth was also studied, since growth was seen as a consequence of nutrition. Physiologists working on the mechanism of plants naturally paid attention to developments in animal physiology. As a result plant physiology was corrected and reformulated in a number of respects, with consequences that were far from negligible. We can make out two distinct themes in the study of functions related to nutrition. The first theme was based on looking at the plant machine in a new way. The usual reason for revising a model is that objections are raised to explanations based on it. We see this when we look at proposed changes in the mechanism of sap circulation. When problems cropped up, there were two possible responses: elements of knowledge already in hand could be recombined in new ways, or new models could be proposed in order to work out novel solutions to the problems at issue.

Geoffroy chose the first course and proposed a new view of the relation between the tracheal system and the circulatory system. A revised version of the circulatory mechanism was needed because observation had shown that sap collected in the roots in liquid form rather than in vapor form as Perrault had asserted. Thus another cause had to be found to explain the motion of the liquids. Malpighi believed that the action of the tracheae caused the sap to rise. By contrast, Geoffroy argued that tracheal action caused the sap to circulate: "The tracheae first swell as a result of the rarefaction of the air inside them and then subside as a result of the condensation of the same air, thus compressing nearby vessels again and again and thereby causing the juices to circulate."[20] Now, it was known that, in animals, nature complements the action of the heart with such accelerative factors as the pulsation of the arteries and, even more important, the pressure exerted on the vessels by certain solids. With this in mind, Duhamel du Monceau found a use for the mechanism suggested by Perrault; the action of the air on plants, he argued, makes up for their lack of a muscular system. The motion of the branches caused by the wind produces "the same effect on the liquor of plants as the action of the muscles on the liquor of animals."[21]

Bazin chose the second course, proposing a new model. Why should a function as essential as circulation depend on the action of the tracheae and thus ultimately on the alternation of hot and cold? If the expansion and contraction of the tracheae were governed by so unreliable a factor, then it was easy to see that the smallest atmospheric disturbance might well interfere with the circulation of the sap, if not interrupt it altogether. This was coupled with a more physical argument. The internal temperature of insects and plants, unlike that of the higher animals, is approximately equal to the temperature of the air. Hence there is no way for the juices to be set in motion. From these considerations it follows that the rarefaction of the air, which is the cause of the sap's motion, cannot be caused by heat, as had been thought first by Malpighi and later by Geoffroy and Duhamel du Monceau. Bazin therefore cast about for another solution. The proper way to determine how the tracheae work was by deduction from their structure. Now, the degree to which a spring can be compressed can be increased by placing it inside a spring-box. Doubtless it was this observation that suggested the following question to Bazin: "Are not the tracheae of insects and plants like the spring-box, that is, a device designed so as to compress the air and increase the degree of thrust?"[22] The tracheae of plants, like those of insects, are conic vessels. The force and speed of the air increase as the size of the duct through which it is flowing decreases. Thus, the compressed air is led into the plant vessels with a force capable of causing the sap to circulate. The same effects are therefore produced by different means. "The internal heat of large animals, by causing the air that they breathe to become more rarefied than the outside air, produces the force necessary for the circulation of their blood; and the tracheae of insects and plants, which inject into their blood an air more condensed than the outside air, bring about the same result."[23] In short, the peculiar mechanism of the tracheae makes up for the lack of heat and constitutes a "principle of motion" not subject to the influence of external factors.

The second theme in the study of plant functions revolved entirely around the uses of the different parts of the plant. Sometimes new functions were added. The earlier nomenclature for the plant parts was generally maintained. There is nothing surprising in this, for animal physiology also attributed several different roles to the same part. Whenever the function of an organ was changed or an additional function was assigned, a new description had to be given of its structure and a fresh look taken at the way in which it functioned. A clear illustration of this process can be seen in the various reworkings of the respiratory function. One possibility, of which Tull availed himself, was to substitute one function for another. Rather than assign the job of cooking the humors to the lungs, Tull assumed that their unique role was to excrete waste products due to excess serosity of the blood. From this point of view, the foliaceous parts contribute not to the preparation of the sap but to its purification: the "leaves are the Parts or Bowels of a Plant, which perform the same Office to Sap, as the Lungs of an Animal do to Blood; that is, they purify or cleanse it of the Recrements, or fuliginous Steams, received in the Circulation."[24] Thus when Tull applied the term "lung" to the leaf, he intended to indicate a similarity of structure rather than an analogy of function, as Malpighi had intended. The structure of leaves and lungs was similar because leaves are "as Lungs inverted, and of a broad and thin Form; their Vesiculae are in Contact with the free open Air, and therefore have no need of Trachea, or Bronchia, nor of Respiration."[25] The difference between Tull and Malpighi is immediately apparent. Malpighi emphasized a mixing process, a process of dissolution of the humors, whereas Tull stressed the unbinding or separation of the components, or, again, to use chemical language, a phenomenon of sublimation.

The other possibility was to add new functions, for example by applying to plants the theory according to which the air plays an active role in the nutrition process. This use of respiration is compatible, moreover, with the purpose previously ascribed to the tracheae. "Today, it is generally agreed that respiration serves two principal ends: one of these is to introduce a few parts of air into the blood, and the other is to facilitate the circulation of the blood by means of the motion of the respiratory organs. Now, all of this applies to plants as well."[26] A choice must be made, however, whether to use the respiration of the higher animals or that of the lower animals as a model. If the former, the first step is to observe that, in the higher animals, the trachea, located near the esophagus, carries the air inhaled through the mouth to the lungs. "We find the same thing in the root, which in plants takes the place of the mouth. In other words, we find two sorts of duct in the root: one receives the juices that serve as food, and these are called the ligneous tubes; the other carries the air into the cells, and these are called the tracheae."[27]

If one follows Bonnet in choosing the lower animals as model, however, the first step is to notice that the tracheae of insects lead to openings in the surface of the skin, the stigmata. Similarly, on the foliaceous parts of the plant, which are in contact with the air, one finds "openings, a kind of *stigmata*, that allow the air to enter into the plant."[28] This statement seems to be confirmed by observation, since the tracheae of plants resemble those of insects, and the large number of tracheae in the leaves tends to support the conjecture. The choice of model is further justified by another consideration. Anyone who wishes to carry out on plants the same experiments that Malpighi and Réaumur conducted with caterpillars has no choice but to identify the pores of the leaves with the stigmata. Insects were known to die when immersed in oil or when even a few drops of oil were applied to their stigmata. In view of the remarkable similarity in the structure of plant and insect tracheae, it was natural to ask whether plants would be able to survive a similar ordeal. To answer this question Bonnet undertook a series of experiments. Just as one might immerse a caterpillar in water or coat it with oil in order to demonstrate the existence of the respiratory function, so one might take branches covered with leaves and immerse them in water or coat them with a substance impermeable to air. The experimental act was thus paralleled by a question that made sense of it, but the outcome remained inconclusive. This was quickly noticed by Bonnet himself. "After further reflection on these various experiments, I have concluded that their results are doubtful. In carrying them out, I failed to take an essential precaution: I did not take care to drive the air away from the outside surfaces of the branches and leaves before immersing them in the water."[29] Nevertheless, many naturalists interpreted Bonnet's experiments as proof of respiration in plants. For Bertholon, "the air that is absorbed in this way by the expiring vessicles and tracheae is later driven away; this alternating cycle constitutes the respiration of the plant.... Immerse several leaves of any kind in a vase filled with water, and soon you will see a rather large number of air bubbles emerging from the leaves and covering their entire surface."[30] For Gesner, too, "the existence of tracheae in the leaves of plants as well as in the stems and roots is demonstrated by the emergence of bubbles on the surface of leaves soaked in water."[31] Finally, the external parts of plants were observed under the microscope. This might appear to indicate a new direction of research, the main concern previously having been to examine the interior rather than the exterior parts. In reality, however, these observations confirmed, clarified, and extended the main points of analogy that had been drawn between the parts of plants and animals, particularly in regard to the cutaneous covering. "The uses of the leaves have much in common with the uses of the human skin. The skin also has its *excretory* ducts, the organs of perspiration. Similarly, it has *absorbent* ducts, which suck in vapors from the surface or its vicinity and carry them to the inner parts of the body."[32] Dissecting the outer portion discloses the diverse membranes of which it is composed. This led H. B. de Saussure to suggest a revised picture of the leaf: "Some botanists, among them some of the most celebrated names in the field, have called the covering of the leaf the 'epidermis,' because they thought it was merely a simple membrane. I have convinced myself, however, that this was a mistake. This envelope I have seen to be a true cortex, having its own epidermis; this is why I have decided to call it the 'cortex.'"[33]

Dissection of the inner portion of the leaf revealed that the cortical structure was dotted with glands. As a result, studies were made of the excretory function, which led to the discovery of imperceptible as well as perceptible forms of perspiration. This distinction was not new, having been proposed earlier by Renéaume: "Plants as well as animals eliminate waste matter in two different ways, as perceptible and imperceptible perspiration. By perceptible perspiration I mean the evacuation through the pores of the leaves of material too coarse to be exhaled and immediately evaporated."[34] Just where on the surface of the leaf these exudations occur had not been made clear, however. Further, they had been confused with the dew that collected on the leaves during the night. It remained for Musschenbroek to clear up these misunderstandings: "After making a number of observations with all due care, I learned that these drops of water are nothing other than the perspiration of plants, which flows continually from their ducts."[35] The droplets of perspiration are observed only in the places where the duct opens onto the surface of the leaf. Guettard observed the same gland network that Saussure had seen; it consisted of bladders of various shapes, usually topped by hairs, threads, or tubes. "It was natural to view these tiny bodies as cutaneous glands, whose function is to allow the imperceptible perspiration to escape. . . . One might look upon the branching ducts that join these bladders as secretory ducts, and upon the hairs or spines as excretory ducts."[36] As to absorption of nutritive matter through the roots, finally, it had been thought that the nutriment made its way into the plant through tiny orifices. But now observation revealed something quite different. "The eye can see this head at the extremity of each fiber only in a clean root, and it is impossible to gaze upon it without being surprised, for what one sees is quite different from what is commonly held concerning the extremities of the fibers. It had been assumed that they were small, to the point of being invisible, and open. On the contrary, as we see them here, they are quite broad but closed."[37]

    The next question to be considered was that of plant growth. Duhamel du Monceau approached the problem in connection with the study of bone growth. In the course of

Jean-Etienne Guettard, plate XXIV from "Mémoire sur la transpiration insensibles des plantes," *Mémoires de l'Académie des Sciences*, Paris (1748). Photo courtesy of the Muséum national d'histoire naturelle, Paris.

NUTRITION

this work he refurbished a theory proposed earlier by Malpighi. The latter had pointed out a similarity between the growth of trees and the growth of bones. He had compared the inner part of the cortex to the periosteum and the ligneous part to the bone. Just as the bone increases in size as the layers of the periosteum harden into bone, so the ligneous body increases in size, Malpighi argued, when circles of cortex are converted into wood. Duhamel du Monceau was all the more willing to adopt this explanation because two lines of research—independent ones moreover—seemed to confirm it. On the one hand, he had shown that bones have a laminated structure. Madder dye could turn bones red in color; if the dye treatments were interrupted, however, they regained their original white. The only possible explanation was that new bony layers were added on top of the existing ones. In addition, Duhamel du Monceau and Buffon had determined what causes eccentricity of the ligneous layers (too much food) and had shown that wood has a laminated structure. "In trees as in animals, the inner layers surrounding the pith are the first to harden, and these are fortified by layers that detach themselves from the cortex in trees and the periosteum in bones, which leads to an increase in size."[38]

THE SYSTEM OF VEGETATION

Another way of looking at questions of nutrition also emerged during the classical era. Rather than select a specific function as a model, Hales began by describing an object that had no relation to the organ centers:

And if we reflect upon the discoveries that have been made in the animal œconomy, we shall find that the most considerable and rational accounts of it have been chiefly owing to the statical examination of their fluids, *viz.*, by enquiring what quantity of fluids, and solids dissolved into fluids, the animal daily takes in for its support and nourishment: And with what force and different rapidities those fluids are carried about in their propper channels, according to the different secretions that are to be made from them: And in what proportion the recrementious fluid fluid is conveyed away, to make room for fresh supplies; and what portion of this recrement nature allots to be carried off, by the several kinds of emunctories and excretory ducts.

Since animals live and grow thanks to absorption, excretion, and the motion of the fluids that carry the nutrients, and since plants also live and grow, argument by analogy suggests that the mechanisms must be the same. "And since in vegetables, their growth and the preservation of their vegetable life is promoted and maintained, as in animals, by the very plentiful and regular motion of their fluids, which are the vehicles ordained by nature, to carry proper nutriment to every part,"[39] the similarity between these two classes of organized beings is clearly a very general phenomenon. This leaves the botanist a wide latitude to accept or reject the idea that plants have organ centers similar to those found in animals. To resolve the question, the first thing to do is to undertake a comparative study of the quantities of nutriment ingested and excreted by plants and humans. Only two outcomes are possible. Either plants and humans (given equal body weights and equal times) ingest and excrete the same amount of fluid, in which case there is no choice but to assume that their economies are similar. Or the two amounts will differ, whereupon the proper thing to do would be to look for structural and functional dissimilarities capable of accounting for the difference. It is legitimate to approach plant life in this way, of course, only to the extent that it is possible to compute the quantity of fluid the plant takes in and eliminates, and, further, only to the extent that such a computation makes it possible to compare reasonably homogeneous pheonomena in the two classes of organized beings.

Consider first the question of calculating quantities. It is clear that, in order to establish the similarity of fluid motion in plants and animals, the procedures by which that motion can be studied must also be taken into consideration. Because the analogy between plants and animals is so "great," "it is therefore reasonable to hope, that in [plants] also, by the same method of inquiry, considerable discoveries may in time be made."[40] Accordingly, the first order of business is to investigate fluid statics, because the sap, like the blood, can be analyzed only in terms of force, velocity, and volume. Hence the methods used by the iatromechanicians must be carried over into plant physiology. The first step was to apply to the plant an experimental device similar to that used to calculate the force and velocity of the blood in the arteries of animals. Since Hales was in complete command of the necessary technique, this procedure posed no major problems. Had he not carried out hemostatic experiments on dogs as early as 1707? Six years later he repeated the same experiments on horses and other animals. He had done these experiments in order to correct results obtained earlier by other researchers. The object was to determine the "force" of the blood in the crural arteries of a dog more precisely than Borelli had done.[41] The next step was to carry out experiments similar to those already performed by certain doctors, involving the use of scales. Santorius, for example, had compared the amount of food ingested with the amount of waste eliminated through perspiration and excretion. The difference represented the weight loss through "imperceptible perspiration." The point of the experiment was to monitor the process by which the animal maintains its weight in balance. Hales's purpose was quite different. By calculating the amounts of food ingested and waste excreted, he hoped to determine the structure of the plant and the manner in which it functions. To accomplish this, it was of course necessary to compare the results obtained from observation of plants with the results obtained from observation of another kind of organism.

This brings us to a second question: Was Hales's choice to use man as a term of comparison a judicious one? In some respects, the choice seems justified. First of all, there were methodological reasons. Using man as an object of experimentation promised to yield more precise results than could be obtained with other animals. Then, too, there was the heuristic advantage that the quantities of material ingested and excreted by man could be related to known features of his anatomy and physiology. In other respects, however, the choice of man is a surprising one. Was it really a good idea to compare plants with man? The question is not unreasonable; Hales himself asked it. But to ask the question is at the same time to answer it. A series of differences between human beings and plants is immediately apparent. Plants, for instance, have neither respiratory organs nor a digestive cavity. This explains why the excrement of animals contains pieces of some size and why animals exhale water vapor through their lungs. The problem then becomes one of restricting the comparison between men and plants to commensurable phenomena. To do this, what more need be done than to deduct from the quantity of food ingested by the human subject the weight of the fecal matter, and from the quantity of other waste eliminated the weight of the exhaled water vapor? It is clear that, once this is done, the structure of the plant corresponds precisely to a well-defined portion of the structure of the human being. Further correspondences then emerge between the constituent elements of the two structures. The quantity of sap taken up by the roots corresponds to the quantity of chyle absorbed by the lacteals. The quantity of liquid eliminated through the leaves corresponds to the quantity of liquid excreted by the kidneys and by the skin through perspiration. Now, "it is found, that seventeen times more enters, bulk for bulk, into the sap vessels of the plant, than into the veins of a man, and goes off in twenty-four hours."[42] This result convinced Hales that the important thing to concentrate on was the heuristic value of the comparison. He therefore noted a series of structural and functional differences between plants and human beings, differences indicated most clearly by the quantitative results of his experiments.

In Hales's work, then, animal structures and functions were destined to serve not as models but as terms of comparison. For one thing, little food is absorbed by the lacteal ducts. But chyle, the product of the first phase of digestion, is produced in the stomach. This being the case, the fact that the roots imbibe a far greater quantity of liquid might have something to do with the comparatively crude state of that liquid for nutritive purposes. "One reason of this greater plenty of fresh fluid, in the vegetable than the animal body, may be, because the fluid which is filtrated thro' the roots immediately from the earth, is not near so full fraighted with nutritive particles as the chyle which enters the lacteals of animals."[43] Second, allowance must be made for the fact that man eliminates half of his excess fluids through the kidneys, the other half being evacuated through the skin, which offers a relatively small area for perspiration. The large amount of material perspired by the sunflower, far greater than that perspired by man, must have something to do with the surface area of the leaves. "And since, compared bulk for bulk, the plant perspires seventeen times more than the man, it was therefore necessary, by giving it an extensive surface, to make a large provision for a plentiful perspiration in the plant, which has no other way of discharging superfluities; whereas there is provision made in man, to carry off above half of what he takes in, by other evacuations."[44] Finally, animals have only a small amount of blood, but the movement of fluids in them is accelerated by the heart. The relatively large amount of sap is no doubt needed to make up for the absence of a central pumping organ. "And the motion of the sap is thereby [that is, by the greater quantity of fluid] much accelerated, which in the heartless vegetable would otherwise be very slow; it having probably only a progressive and not a circulating motion, as in animals."[45] The large amount of sap therefore helps to accelerate its flow. Hales expressed the fluid pressure in the sap ducts in terms of multiples of the blood pressure in the crural arteries of certain animals rather than in inches of mercury. In a vinestock, for example, the pressure is almost five times as high as the blood pressure in the crural artery of a dog, and eight times as high as the blood pressure in the crural artery of a deer. To sum up then, Hales was able, by comparing plants with human beings, to discover differences that pointed to what is specific about the economy of plants. The roots absorb copious quantities of raw nutrient, the leaves provide broad surfaces for transpiration, and the sap has a progressive rather than a circulating motion. Broadly speaking, this is the "system of vegetation."

What Hales was proposing was not so much a theory as a working hypothesis. Experiments were begun along several lines in order to determine whether or not plant parts do in fact play the role assigned to them. The first problem was to establish that the food absorbed by plants is indeed still raw nutriment rather than a complete food like chyle. Mention must be made here of Woodward's work, which was in fact cited by Hales. Woodward experimented on a particular kind of object, three sprigs of mint of the same species and weight. Each sprig was placed inside a hermetically sealed flask, the first in spring water, the second in rain water, the third in river water. After a certain period of time, Woodward compared the amount of water consumed and the weight gained by each plant. Thus the ratio of the weight increase to the amount of water imbibed could easily be determined. The results showed that the more nutritive matter there was in the water, the greater the weight gained by the sprig of mint: "The plant is more or less nourished and augmented, in proportion as the water in which it stands contains a greater or less quantity of proper terrestrial matter in it."[46] Another experiment, in reality just another version of the preceding one, gave similar results. In this experiment Woodward used irrigation water, water mixed with common soil, and water mixed with compost. If the nutritive material is carried in the water, the soil must contain large quantities of it. This suggested to Hales the idea of measuring nature's reservoirs. The procedure was simple. All that had to be done was to take equivalent volumes of soil at various depths, weigh them, let them dry, weigh them again, and calculate the difference between the two weights thus obtained for each sample.

That leaves are organs of perspiration could also be checked by means of very simple experimental measurements. All that had to be done, in fact, was to relate the amounts of liquid taken up and perspired by the plant to either the surface area of the leaves or the intensity of the ambient heat. By cutting off a few leaves, one could vary the surface available for perspiration and thus verify that exhalation is proportional to the surface area of the foliaceous parts of the plants. For example, when Hales placed a leafless branch in one vase and a leaf-bearing branch of the same species in another vase, he found that the first branch took up only one ounce of water, whereas the second took up twenty to thirty ounces. By examining the effects of temperature variation, he was also able to demonstrate the influence of heat on perspiration. Hales found that the weight of the plant decreases when the heat increases, whereas the weight increases when the heat decreases. To follow the movement of the sap, which sometimes rises and sometimes moves in the opposite direction, a glass tube containing mercury was fitted to the top of one branch.

NUTRITION

Finally, in regard to the nature of the sap's motion, Hales followed Harvey's procedure in one respect, basing his argument on quantitative measurements. In other respects, however, Hales took a different course. Harvey had worked out a theory of the circulation of the blood. Hales's theory was one of oscillation of the sap, of a back and forth motion. Harvey substituted the concept of circulation for the concept of irrigation, which was based on the assumption that the liver produces large quantities of blood. His method was to calculate the amount of new blood the heart would have to deliver to the other parts of the organism if the irrigation hypothesis were true. The result he obtained was that in one hour the heart would have to pump three times the body's weight in new blood. Since this was absurd, Harvey conceived the idea that some of the blood must return to the heart, which led to the idea of circulation. Hales's procedure was the reverse of this. If we assume that the sap circulates, it follows that the plant needs to take in only small amounts of matter. Hales's calculations showed that, on the contrary, plants take in large amounts of nutrient. It therefore seemed absurd to maintain that the sap circulates: "What prodigious velocity the sap must have if this moisture, or at any rate the bulk of it, must rise to the top of the tree, then descend, and finally climb again before being exhaled through perspiration."[47] In short, Harvey emphasized the conservation of the blood (the amount of food absorbed and eliminated being small), whereas Hales emphasized the dissipation of the sap (the amount of sap imbibed and eliminated being large). There is a further point to be made. Just as Harvey had invoked the old analogy between the microcosm and the macrocosm, so Hales compared the sun to the heart. By substituting a natural context for a magical one, however, the analogy was made more specific: the sun is not so much the heart of the world as of plants. The arteries of a dying dog are similar to the vessels of a severed branch—are they not?—in that both are compressed to such an extent as to impede the passage of blood in the one case, sap in the other. Both are deprived of the "principle" that makes them work. Further, since the blood is forced through the veins by the expansion and contraction of the heart, it is quite probable that the sap, too, "is carried up to great heights . . . by the vigorous undulations of the sun's warmth, which may reciprocally cause vibrations in the vessicles and sap vessels, and thereby make them dilate and contract a little."[48]

Throughout the eighteenth century research on this function continued along the path blazed by Hales. Lacking a chemistry to analyze the nature of nutrition and of the nutriments of plants, however, researchers had to concentrate instead on the role played by the leaves and the path followed by the sap, as well as on plant growth. Research on leaves was of two kinds. The first revolved around challenges to Hales's idea that the leaves absorb moisture. According to Hales, a drop in the temperature was enough to cause the leaves to absorb moisture from the air. If trees do in fact take in moisture from the air through the leaves, however, then a tree whose leaves are not exposed to the open air should absorb less moisture and therefore perspire less than a tree whose branches are uncovered. This observation no doubt inspired Guettard to try a very simple experiment. He sealed all the branches of one orange tree inside a glass sphere. Two branches of two other trees were also enclosed, while the remaining branches were left in contact with the air. Only two outcomes were possible. Either the trees would draw in moisture from the air, in which case the tree completely encased in glass should perspire less than the trees with only two of their branches enclosed, or else the perspiration would be equal, which would refute Hales's contention. Now, the experiment showed "that a tree whose branches are all completely enclosed in a space not open to the air does not suffer at all, at least not in any perceptible way, from this artificial state, so that the leaves may not draw from the air as much moisture as is commonly believed."[49]

Other research on leaves revolved around perspiration. The first order of business was to measure the amounts perspired more precisely, not in order to verify the existence of perspiration (Hales had already done that) but merely to refine his results. It was necessary to redo his experiments in such a way as to avoid the problems inherent in his procedure. Guettard's experimental methods were clearly an improvement over those of his predecessor. It was therefore natural that he should be critical of Hales's results as well as of his technique. "In adapting Mr. Hales's procedure, I had to devote some thought to avoiding certain problems inherent in the method. These problems were of little moment to Mr. Hales, in view of the purposes he had in mind, but for my own purposes I needed to take precautions in order to avoid them."[50] In other words, Hales's experiments were poorly designed, and the conditions under which they were carried out were too close to the conditions prevailing in nature. To study perspiration he had used a simple retort. But in the retort itself an evaporation-condensation cycle occurred. Furthermore, the branch in the retort would have reabsorbed part of its own liquid. The heat of the sun acting on the liquid maintained part of it in the vapor phase. To avoid these problems, Guettard used an ingenious device, a glass sphere one foot in diameter, which, in addition to the usual opening (located on one side, through which the branch was introduced), was fitted with a neck on its lower portion. This neck mated tightly with the mouth of a bottle that was completely buried in the earth, so that the liquid perspired by the plant might collect in the bottle. This made it possible to measure the amount of liquid perspired more precisely than Hales had done.

Another tack was to consider how perspiration was

affected by still other factors. Botanists interested in studying the effects of the environment on plants naturally turned their attention to factors likely to affect perspiration, thus opening a new avenue of research. Apart from the heat, Guettard observed that other factors, such as light, were important. A branch placed in a sphere sheltered from the sunlight but maintained at a temperature higher than the ambient temperature perspires less than a similar branch enclosed in a sphere placed in direct sunlight. "It thus appears that the immediate action of the sun's rays increases perspiration, and that one plant may be in warmer air than another plant and yet perspire much less."[51] There was no question here of a chemical process, however. Guettard believed that the whiteness of plants grown in darkness was due to the absence of light only insofar as light affects perspiration, as the following citation shows: "I believe that I have found the true cause of this phenomenon in the arrest of perspiration: the large quantity of water that gathers in the parenchymatic vessicles of these plants inflates them, causing them and therefore the plants to lengthen, and perhaps this is one cause of their turning white, since once these plants are again placed in the open air and exposed to the sun, they regain their ordinary color."[52] Electricity as well as heat influenced perspiration. "Plants are analogous to the instruments of physics in that they exhibit identical phenomena to us, because their motions are governed by the composition of the atmosphere. These motions are accelerated when the atmosphere is charged with positive electricity; and reversed . . . when the electricity of the atmosphere is negative."[53] Finally, it was possible to show that the amount of nutriment imbibed by the plant varies with the age of the foliaceous parts and the intensity of the heat but does not depend on the amount of water taken in. If the experiment is extended over a three-month period, it is found that perspiration varies with the intensity of heat but not with the amount of rainfall. Between June and August the amount of rainfall decreases, but the intensity of light increases. Perspiration, however, is greater in August than in June. "It follows that it is not the quantity of water supplied to the plant that determines the amount of perspiration. . . . In general, then, it may be said that the perspiration of plants is due only to the more or less constant action of the sun."[54]

The question of the path followed by the sap also received a fresh look. This was because there were different schools of thought on the subject. Those who agreed with Hales that heat is the principal agent in the motion of plant fluids were obliged to take the view that the sap rises through the cortex, for the cortex, like the leaves, is exposed to the heat. In contrast, Renéaume and Parent had based their systems on observation. Renéaume had noticed that hollow trees continue to grow, so that the plant nutrients must be carried by vessels in the cortex. "The cortex contains and enwraps the vessels that carry the nutritive juice to all parts of the tree."[55] Parent, however, had seen an elm that continued to live though stripped of its bark, and he therefore held that the nutritive sap is carried by the ligneous tubes. "The platan and cork oak shed their skin and grow a new one, like serpents. During the transition they are not fed by the bark, hence they never are."[56] If the mechanism of fluid transmission is oversimplified, as in the case of Hales, the end result is that the only structural distinction made is between wood and cortex. If, however, the theory is based on examination of special cases, as in the cases of Renéaume and Parent, there are always counterexamples to stand in the way of generalization. In the end, the problem was similar to that faced earlier by anatomists, for the vascular system of animals was also difficult to understand. There, however, injections had proved a successful method; why not use injections in the study of plants? To settle the question of the path followed by the sap, plants were made to absorb colored liquids. Sarrabat soaked the roots of various plants in water colored red by the juice of the phytolacca fruit. This experiment revealed that "the vessels intended to carry food into the body of the plant are neither in the pith nor in the bark nor between the bark and the wood, but in the ligneous substance of the plant; or, to speak with greater precision, we should say that these vessels are actually ligneous fibers contained between the pith and the cortex of the plant."[57] The comparable procedure in animal anatomy was probably experimentation with the ingestion of different foods rather than injections.

This latter point raises a problem, however. Since we know that when animals are fed red dye, only the bony parts turn red, it is clear that Sarrabat's experiments do not prove what he thought they proved. For, in animals, the dye passes through the soft parts and lodges only in the hard parts; the same might be true in plants. Another problem was that plant saps are usually similar in color to the liquids used to test their absorptive characteristics. Bonnet came up with a way to get around these difficulties: he changed the plant used in the experiment, substituting for the tree a herbaceous plant that was etiolated, or white in color. He reasoned that if plant parts do absorb color in the same way as animal parts, the etiolated beans should not be colored by the various dyes they were fed. Indeed, because the parts of the bean were softened by etiolation, they should be even less capable of absorbing the dye. In fact, however, the beans did become colored, just beneath the surface, and Bonnet observed "lines of the most beautiful black, as clear, as straight, and as neatly terminated as if they had been drawn with pen and rule."[58]

Nutrition is responsible not only for the sustenance of the various parts of the plant but also for their growth. Thus plant growth became a related area of research. The first step in studying increase in length was to establish, as Hales did, that the different parts of the plant grow unevenly, or, rather, that the tenderest parts are the ones that grow and that the amount of growth diminishes as the hardness of the stem increases. Hales split a vinestock lengthwise and made marks every quarter of an inch along its length. After a certain lapse of time, when the size of the stock had increased, he noticed that the space between marks in the softest part of the wood was three inches in length. The spacing in the harder portions of the plant remained the same, however. Duhamel du Monceau repeated Hales's experiments and came up with the same results: "What are we to conclude from this experiment? That, as long as all the parts of the stem are herbaceous, the stem grows over its whole length, but that this property of extension diminishes in proportion as the ligneous body forms inside the plant, and growth ceases entirely when the ligneous body is fully formed."[59] Hales and Duhamel du Monceau differed as to the interpretation of growth in breadth. For Hales, the wood grows only by increase in length: "These considerations make it not unreasonable to think, that the second and following years' additional ringlets of wood are not formed by merely horizontal dilatation of the vessels; for it is not easy to conceive, how longitudinal fibres and tubular sap-vessels should thus be formed; but rather by the shooting of the longitudinal fibres lengthways under the bark as young fibrous shoots of roots do, in the solid Earth."[60]

Duhamel du Monceau, on the other hand, maintained that the ligneous layers grow from the outside in. He viewed the question as related to the problem of scar formation in injured trees, in the light of a theory proposed earlier by Grew and revised by Duhamel for the present purpose.[61] Wounds in the flesh close and scars form owing to the growth of a cambium; similarly, wounds in trees close and ligneous layers form in trees as a result of the organization of a cellular tissue. In short, the ligneous layers are a fresh production of the cortex. Duhamel du Monceau carried out a careful experiment that seemed to verify this theory of growth. He removed a ring of bark, cemented a tube around this part of the trunk, and filled it with water. Two things might happen. Either the mucilage would dissolve in the water, indicating that it was not "organized" matter. Or a scar would form over the wound, showing that the mucilage was turning into wood, hence that it was organized. "These experiments, which I should have liked to repeat, convinced me that I was correct in thinking that this apparently gelatinous substance was in fact organized; if it had merely been a mucilage, it would have dissolved in the water." [62] Hence it followed that the ligneous layers came from the cortex.

Finally, it is important to notice the differences between the system of vegetation as worked out by Hales and clarified and corrected by Guettard and Duhamel du Monceau, on the one hand, and plant mechanism on the other hand. It would be a mistake, however, to draw too hard and fast a line between these two ways of understanding vegetable economy. Plant mechanism, which was worked out for the most part in the late seventeenth century, preceded the system of vegetation. Hales therefore had his attention focused on the circulation of the sap, respiration, and growth. His problem was not only to criticize the work of the circulationists but also to suggest new ways of interpreting their results in the light of the theory of alternating motion of the sap. Perrault, for example, had made much of defoliation. The counterpart of this procedure in animals was not desquamation or amputation but rather bleeding white, given the fact that the leaves are traversed by a network of veins. Thus the effect of defoliation is to deprive the plant of a humor analogous to venous blood. Hales did not deny that removing a plant's leaves kills it, but it dies, he argued, because its organs of perspiration have been removed, not because its circulation has been interrupted. "Upon the whole, I think we have . . . sufficient ground to believe that there is no circulation of the sap in vegetables;

notwithstanding many ingenious persons have been induced to think there was, from several curious observations and experiments, which evidently prove, that the sap does in some measure recede from the top towards the lower parts of plants, whence they were with good probability of reason induced to think that the sap circulated."[63] New experiments were done to refute the theory of circulation. Mustel experimented on two rose bushes. One was in a greenhouse with its branches outside; the other was planted in the open air, but its branches extended inside the greenhouse. Now, only the parts of the rosebushes inside the greenhouse produced buds, flowers, and fruits. It followed from this that "the existence of circulation of sap in vegetables, like the circulation of blood in animals, is a purely imaginary system devoid of any reality."[64] Plant respiration, for its part, could be investigated with the aid of the vacuum pump, which made it possible to prove that plants contain air. In this there was nothing very new, since Mayow had already demonstrated that plants cannot live very long without air.[65] Hales conducted the following experiment. A branch was soaked in water and placed in a pneumatic machine while the stem remained in the open air; when the air was pumped from the bell jar, Hales saw bubbles emerge. This clearly proved that air was passing through the plant. The "observation corroborates Dr. Grew's and Malpighi's opinion, that they are air vessels."[66]

Last, the problem of the growth of the wood was reexamined. Research showed that the ligneous layers do indeed grow out of the cortex. One way of doing this was to graft a peach bud onto a plum tree; another was to insert a sheet of tin between cortex and wood. The first experiment led Duhamel du Monceau to say that thin layers of wood are produced by the bast contained in the graft, because the layer that grew was of peach wood rather than plum wood, as could be seen from its color. The second experiment showed that the new wood that formed on the layer of tin grew out of the cortex. Still the problem of where the ligneous layers come from remained unresolved. The question was whether the layers of bast harden, as Malpighi believed, or whether the wood is produced by the cortex without having been a part of it, as Grew maintained. To answer this question, Duhamel du Monceau carried out a careful experiment. He believed that if he inserted a silver needle into the cortex, only two things could happen. Either the needle would remain in place and the layers of the cortex would continue to be part of the cortex, which would show the correctness of Grew's opinion that the wood is a product of the cortex. Or the silver wire would become encased within the wood, which would prove Malpighi's theory that the cortical layers turn into wood. Unfortunately the experiment did not actually resolve the conflict between the two theories. When the needle was inserted beneath the epidermis of the cortex, that is, in its outer portion, it remained embedded there. The result was the same when the needle was stuck into the middle of the cortex. When it was inserted into the innermost layers of the bast, however, it became encased in the wood. Thus no conclusion could be drawn. "These observations are very favorable to Malpighi's opinion, but they do not contradict Grew's; for it is incontestable that the ligneous layers are produced by the cortex, so that they cannot acquire all their hardness in one stroke."[67] It was not even possible to reject Hales's theory, because a tree stripped of its bark regenerates a new bark: "The wood stripped of its bark can, therefore, produce a new bark, beneath which ligneous layers are formed, which is largely consistent with Hales's opinion."[68]

In the end, then, the theory of the circulation of the sap was refuted and the theory of respiration confirmed. The problem of growth, however, was not resolved: "I confess that my researches and experiments have not led me to a complete solution of the problem on which I was working, inasmuch as I have not yet decided whether Malpighi's opinion is to be preferred to Grew's and do not know how to explain several peculiar facts turned up in the course of my experiments."[69]

THE TWO KINGDOMS, PLANT AND ANIMAL: SIMILARITY AND DIFFERENCE

When the physiologists ended their work on nutrition, one problem had yet to be resolved. To conceive of the economy of vegetables in terms of nutrition was to identify a vital phenomenon common to both plants and animals. The theory of nutrition therefore had to satisfy two requirements. Since nutrition is a function common to both classes of organized beings, there must be some similarity between the way it is accomplished in plants and the way it is accomplished in animals. But there must also be some difference between the two, for otherwise there would be identity rather than kinship. The nature of the similarity was determined by the way one went about establishing the theory of nutrition in plants, and this then also determined the nature of the difference. It turned out that the proponents of plant mechanism ran into insurmountable difficulties when it came to elucidating these similarities and differences.

Basically, the similarity between plants and animals was supposed to be based on the circulatory and respiratory functions, whose configurations were said to be similar. The leaves did the work of the lungs, the juices circulated, and the roots furnished sustenance to the sap just as the lacteals furnished sustenance to the blood. Thus plants and animals were held to be identical in some degree. The common functions, it was argued, are organically expressed in both, but by different organs. Thus "the differences peculiar to, and characteristic of, plants, are these: they cling intimately to the soil from which they are born, and they lack digestive organs. In plants we find nothing that corresponds to the mouth, the stomach, or the intestines of animals."[70] Since plants lack masticatory and digestive functions, they are indeed different from animals. Accordingly the similarity between the two classes of organized beings is incomplete. When it came time to specify what the resemblances were, however, there was no choice but to say that with respect to nutrition plants and animals are completely identical, thus eliminating the distinctive differences from view. Plant mechanists were trapped by this strict logic of inclusion and exclusion. Both possible ways to overcome the dilemma ultimately proved to be of no avail.

The first thing they tried was to develop the analogies between the plant and animal kingdoms. The results were hardly satisfactory, however, and it made no difference whether one started with similarities or with differences. Duhamel du Monceau started with differences, noting that plants are distinguished from animals by the absence of a digestive cavity: "The roots, which may be compared to the lacteals of animals, take up the juices that nourish the plant from the earth." With the difference between the two kingdoms assured by this analogy, one could then go on to bring out the similarities by developing its implications. If the roots are analogous to the lacteals, then the preparation of the sap must take place in the soil: "The earth is, therefore, in some ways, the stomach in which the plant's nutritive juices are digested."[71] The missing functions (digestion and mastication) thus reappear, as it were, in the background of the picture, thereby reestablishing the similarity between plants and animals. Sarrabat instead began with similarities. Plants, like animals, have a digestive function because the nutrient taken up from the earth is conveyed to "the principal stomach of the plant, at the node or insertion of the stem onto the root." The next step in Sarrabat's argument was to specify the differences between plants and animals. He observed that the root is covered with a large number of orifices. Thus the plant ingests its nutrient "as it were, through hundreds of mouths, where it receives an initial digestion similar to that which the animal's food receives from mastication." There is a further difference: "The animal's entrails are folded back on themselves but do not divide, whereas those of the plant divide into several branches and terminate in an infinity of imperceptible orifices through which excrement is eliminated."[72] In the end all these ingenious devices proved fruitless, however. If a quasi-organic link is established between the plant and the soil, as Duhamel du Monceau believed, the end result is to animalize the earth. But if one multiplies the number of organs of nutrition, as Sarrabat did, the result is to over-animalize the plant.

Another way out of the dilemma was to borrow a different model from animal physiology, the model of glandular secretion. There were two apparent advantages to using this model. First, the function involved was quite clearly defined, which made it possible to specify the differences between plants and animals. Second, it was a function under which the complex mechanisms of animal nutrition could be subsumed, since "nutrition is another branch of secretion."[73] In other words, it was possible to bring out the similarities between plants and animals at the same time as the differences. This was possible because nutrition was also defined as a function of quite general significance, involving the preparation of specific nutrients for various parts of the body. The question could be approached in two ways, however, depending on which theory of secretion was invoked. One possibility was to use the Newtonian model, as Parsons and Bradley did. Once plants were identified with glands, the differences between plants and animals at once became clear: the plant is composed of various kinds of vessels; so simple a structure contrasts with the more complex structure of an animal, in which there are such specific organs as stomach, kidneys, and circulatory system. The similarities between plants and animals cannot be expressed in terms of analogies between their parts without risk of running into a contradiction. If, for example, the root were said to be analogous to the mouth, it would be impossible to view the plant as having the structure of a gland. Once again, the problem of what the difference between plants and animals is would arise. Actually, however, the difficulty is illusory. Not only is similarity compatible with difference but, curiously enough, similarity derives from difference. The plant being a gland, the absorption of the nutrient through the roots must be a secretion. Furthermore, since attraction indicates a sort of inclination or preference, a match between the secreting organ and the liquid secreted, there is reason to consider the root as exercising powers of selection. "The root, that is, the principal part of it, receives into it such juices of the earth as are proper for it, and no other."[74] At last, the similarity between the plant and the animal becomes clear: Plants, like animals, select specific foods.

The other possibility was to adopt a Cartesian model, as Mariotte and Tull did. The plant having been identified with a gland and animal nutrition with a secretion, similarity is reflected in structural analogy. "The aliment coming from the earth is not suitable to nourish the main parts of the plant; but, in light of the analogy with vegetation in animals, the nutrient must be improved by passing through variously structured tubes, just as the blood is improved by passing through the vessels of the lung, the liver, and various glands."[75] Difference must be compatible with similarity. Thus the analogy appropriate to the difference in question cannot be a structural one. If the roots were held to be analogous to the lacteals, the similarity would in fact vanish because there would no longer be any digestive function in the plant. The difficulty is illusory, however, because difference is entirely compatible with similarity. More than that: difference derives from similarity. Since the plant is like a gland, the ingestion of food by the roots must be a secretion, a sifting or filtering operation. Thus plants can absorb any kind of food, provided the particles of nutrient are no larger than the pores in the roots. The question was looked at in this way by Mariotte, who concluded, "Since plants indifferently take up whatever is dissolved in the water touching their roots, it follows that there cannot be principles that one plant attracts while others do not."[76] In other words, plants, unlike animals, do not choose their aliments. Ultimately, then, Newtonians and Cartesians arrived at opposite results. The opposing positions led to divergent conclusions.

They both drew conclusions with regard to the activity of plants, which meant that prominence was given to the similarities between plants and animals. The focus of interest shifted now from the plant to its food. There were two ways to look at the nature of plant nutriments. If one held, with the Cartesians, that plants do not choose their foods, then it followed that the same substances nourish all species: "Plants of the most different Nature feed on the same Sort of Food."[77] If, on the other hand, one followed the Newtonians in maintaining that plants select their food, one was forced to defend the contrary opinion: "There are as many distinct qualities in the earth as there are different kinds of plants, and each plant draws from the earth only those spirits which are suited to it."[78] The Newtonian and Cartesian explanations did share one assumption in common, namely that the nutritive particles taken up by plants were to be found in the soil, ready made. The only question was whether the substances of the various humors were contained in the earth materially or formally, just as in animal secretion the "only question is whether the substances of these humors ... are contained materially or formally, as one says, in the blood."[79] An answer—at any rate an implicit one—had already been given to this question, since the Newtonians maintained that plant fluids exist in the earth only as forms requiring specification. Thus, "the Earth is their natural *Matrix*,"[80] matrix being used in its technical sense, meaning mold. The Cartesians, on the other hand, contended that the humors exist materially in the earth, because the function of the mold is fulfilled by the plant: "What gives plants their different qualities, then, is the more or less precise separation or union of principles, and the different proportions thereof produced by various filtrations and divisions through the pores of the different structures."[81]

Cartesians and Newtonians were both mistaken as to the significance of their respective conclusions. This led to criticisms of earlier assertions, which were revived for the occasion. It also gave rise to a series of observations and experiments intended to corroborate one thesis or the other. Consider the criticisms first. In the view of the Cartesians, the claim that the earth contains different foods for each species meant that plants played no active role in the nutrition process. The plant being completely passive, its growth must be due to mere accretion of matter, a sort of crystallization. Accordingly Bradley and Parsons were viewed as mere imitators of Aristotle. For what had Aristotle contended, if not that the different qualities of each part of the plant were to be found ready-made in the earth?[82] "We are convinced, that 'tis the Vessels of Plants that make the different Flavours; because there is none of these Flavours in the Earth of which they are made, until that has enter'd and been alter'd by the vegetable Vessels."[83] In the view of the Newtonians, on the other hand, the claim that all plants absorb the same food meant that nutrition took place passively within the plant; their view was that if the food was the same, the nutriment must be homogeneous. The plant was therefore supposed to be the result of some sort of transmutation. On the whole, Mariotte and Tull were rather close to the position of the alchemists. Had not van Helmont maintained that water, a homogeneous element, is the principle of plants?[84] "No Conception can be had of a Power, in so small an Organization as a Seed or Bud contains, to transmute a simple Fluid into so many different Substances, as even one Plant contains."[85] Next, consider the observations and experiments that tended to corroborate these opposing theories. First of all, there was the fact that some plants grow in distilled water. The Cartesians maintained that this element (the water) contains a plethora of elementary particles, which the organs of the plant are able to recombine in various ways. The Newtonians argued instead that, because every plant has the power of choice, each one absorbs only the particles appropriate to it. Another line of argument revolved around crop rotation. Obviously, the meaning of rotation was interpreted differently depending on which theory one happened to believe in. Tull maintained that different plants forage for their food at different depths, whereas Parsons argued that each species exhausted its own particular nutriment.

The other set of conclusions pertained to the similarities and differences between plants and animals with respect to nutrition. Contradictions emerged, however, that had been concealed for a time by the use of the secretion model. The Newtonians argued that plants, like animals, select their food. It was natural, then, that they should compare the roots to the organ in the animal that possesses the power of choice. Bradley offered the following argument in this connection: "The root, which is like the mouth of the tree, draws the sap or juices out of the earth and conveys them to the trunk, where they enter into the cavities or esophagi and entrails of the tree."[86] Conversely the Cartesians argued that plants, unlike animals, do not choose their nutriment. Accordingly they were forced to compare the roots with the parts of the animal that absorb the food mechanically or, as it were, blindly. It was from this standpoint that Tull argued that "the Mouths of Plants, situate in the convex Superficies of Roots, are analogous to the

Lacteals, or Mouths, in the concave Superficies of the Intestines of Animals. These spongy Superficies of animal Guts, and vegetable Roots, have no more Taste or Power of refusing whatever comes in Contact with them, the one than the other."[87] Of course, the point of these analogies was to make explicit the nature of the choices being made by the plant, which were obviously related to the theories of secretion put forward by animal physiology. That said, it is clear that Bradley neglects the differences and Tull neglects the similarities.

By contrast, specifying the nature of the similarities and differences between plants and animals with regard to nutrition presented no problem for the theorists committed to the system of vegetation. In the first place, the notion of nutrition was very broadly defined: Nutrition is a faculty that works to preserve the individual by constantly repairing losses due to the inherent tendency toward decomposition. Repairs were accomplished by assimilating new substance. With such a definition, it was natural to look at the different ways in which this might be done. Animals have a series of complex organs for the purpose, whereas plants have a simpler structure. The sap, in fact, enters the roots in the form of a vapor. The leaves act as a pump, which raises the food particles. In between, the trunk works like a thermometer, for the alternating motion of the sap depends on the temperature. Hence Mustel maintained that "there is no similarity whatsoever between the two kingdoms, plants having no parts that correspond in structure or function to the parts of animals: we find nothing in plants resembling the heart, the lung, the stomach, etc., all of which are found in animals. The structure of plants is very simple, very uniform; the ligneous and cortical fibers, the vessicles and vessels proper are the only components of their visceral systems, and these viscera are distributed throughout the body of the plant."[88] Although the anatomical disparity between plants and animals was thus clear, it was nevertheless obvious that a nutritive function did exist in plants. "The difference between animals and vegetables therefore cannot be based on the way they nourish themselves."[89]

Instead, the difference was based on the contrast between the complexity of the nutritive system in animals and its simplicity in plants. It was argued, in effect, that because animals have respiratory and circulatory systems, their structure must allow for mobility and sensitivity. By contrast, the simplicity of plant structure made this kind of organization superfluous. "Nature's great aim in vegetables [is] only that the vegetable life be carried on and maintained."[90] Thus in plants there is no animal life. The difference between the plant kingdom and the animal kingdom is based on the different ways in which nutrition is accomplished in each. As a result, the similarities between the two kingdoms in regard to this important function are left untouched.

BOTANY AND THE METAPHYSICS OF LIFE

The study of plants, like the study of animals, was not wholly isolated from metaphysics. Descartes had maintained that God created matter, established the laws of motion, and thereafter ceased his intervention. Science therefore has no need to go back to God. "Thus we shall lose no time investigating the ends that God set himself in creating the world, entirely banishing from our philosophy any search for final causes; for we must not presume to think that God wished to take us into his confidence."[91] If no intelligence governs the existence of things, however, may it not be chance that governs instead? Materialist atheism provoked a reaction on the part of those who sought to uphold the Christian tradition; by contemplating final causes, it was held, one could arrive at proof of the divine wisdom. Besides, as everyone knows, machines are built for a particular purpose, and no machine can be understood without reference to its builder's intentions. Thus, as François Jacob has pointed out, "the attempts made during the classical age either to accentuate or limit mechanism sprang less from an attitude to contemporary science than from metaphysics."[92] We see this clearly in the divergent interpretations that were given of plant growth and vegetation.

Some were motivated by a desire to restrain the ambitions of mechanism. They felt that matter and motion alone were not enough to produce and preserve organisms as complex and as perfect as plants or animals. A Cartesian may well be able to explain terrestrial and celestial phenomena in such a way as to require only matter in motion. By contrast, "He will never succeed in producing a similar explanation for plants and animals, since he is incapable of urging anything probable as to their ultimate origins; inasmuch as it is manifestly clear that such objects could never have been the result of the disorderly and fortuitous motion of corpuscles."[93] What was to be done but to assign to God prior possession of the qualities of wisdom, goodness, and foresight, which are characteristic of man? Hence there was a tendency to anthropomorphism.

Let a Pyrrhonist again examine in as many ways as he may wish the earth and the rainwater, of which, as we have demonstrated, most plants are composed, and let him then tell us whether he is capable of proving in that way how it happens that, when one sows the seed of a beautiful flower, or the seed of some poisonous plant, in the same soil, each produces a plant of its own species, whose shape, virtues, and properties are so different; and let him say to us, whether it appears to him in any way probable, that all that can be done without any wisdom. . . . And if the thing is impossible for him (as it has been until now), should he not recognize in these phenomena a wisdom infinitely superior to the wisdom of men.[94]

To hold that plants are formed by God, however, is to attribute to him a function incompatible with his dignity and, when all is said and done, to degrade him. Ray maintained that it was more reasonable to invoke a "plastic nature," as Cudworth did. Instead of God, this plastic nature would accomplish the providential work of moving matter in a regular and ordered fashion. The intelligence that presides over the formation and conservation of plants is an immaterial substance distinct from bodies, a substance that is to God what the artisan is to the architect. In other words, this intelligence is an instrument, a docile executant that obeys the will of the creator. "In order to do all of this and give each part its just proportions, what seems to be required is an intelligent *plastic nature*, capable of knowing and regulating the entire economy of the plant."[95] The dignity of God is thereby restored, and plant bodies are not the result of chance.

By contrast, the second attitude to plant production and vegetation was based on a way of looking at things that invoked neither the perfection of divinity nor divine providence. The different qualities of plants were said to be produced by the motion of the parts of matter, that is, by the different combinations of the principles that make up the bodies of plants. On this basis, it was possible to investigate, as Mariotte did, the conditions under which these filtrations preside over the emergence of the qualities of a particular body. Filtering of this sort must occur in combustible rocks, for otherwise the inflammable matter could not be separated from the aqueous moisture that would extinguish the fire. It also occurred in precious stones, for filtration was the way the purest and most transparent parts were isolated. Finally filtration governed the formation of metals. This, then, explains why "we believe that there is nothing so fortuitous as the production of plants; that their vegetation differs but little from that of stones and metals; in a word, that the best organized plant is but a simple and easy effect of the general movement of matter."[96] Experiments can be carried out to show that the differences between plants result only from the arrangement of salts, earths, and oils. For Mariotte, it was enough to take a pot with some earth in it; since one could plant three or four thousand different plants in the pot, obviously each species did no more than rearrange the same food. One could also graft a branch of a *bon chrétien* pear tree onto a wild pear tree; the same sap that would have produced small, bad-tasting pears would, once it entered into the grafted branches, yield large pears of pleasant flavor. Thus the upshot of these experiments is that the observed differences between plants "result solely from the more or less perfect union of some of these crude principles and their simplest parts or separations."[97] Consequently the growth of plants can be studied without regard to the perfection of the divine. "Those who embrace our opinion can boast of being strict Cartesians, whereas those who allow for a providence of God in the production of plants, other than the general motion of matter, are lax Cartesians who have abandoned the rule of their master."[98]

The Cartesian theory, however, is inextricably bound up with a metaphysics; the identification of animals as automata is the counterpart of the identification of the soul with thought. Two further arguments, however, helped fill the unbreachable abyss that Descartes had placed between animals and man. For one thing, if it is true that a simple mechanism governs the most complex animal functions and marvels of instinct, then human actions can be explained in the same way. For another, if the soul begins with sensation, then there is no reason to deny one to animals. Would they give signs of feeling if they had none? In short, men and animals are reduced, in either case, to the same level. The upshot of these two extreme attitudes is irreligion. Nevertheless, the eighteenth century was prepared to maintain the Cartesian distinction. What more was required than to show that the modalities of alimentation in the animal demand a mobile and sensitive structure? In other words, feeling and movement, far from indicating the existence of a soul, are determined by need. Knowledge of plant life turns out to be very useful in this perspective, because it makes possible an explanation of animal life in terms of plants. Because of this, the spiritualist implications of Cartesian theory were rediscovered, and additional material was brought forward in order to rework the argument. In the first place, plant mechanists made digestion the criterion of animality. In plants, there is respiration, a kind of circulation, and the roots bring nutritive matter to the sap just as the lacteals do to the blood. "Thus plants apparently draw their nutritive juice through outside roots, whereas animals draw theirs through roots inside their bodies."[99] For Boerhaave, then, the animal was a plant turned inside out. To view the beast as a vegetable was to lower its status, however, while at the same time raising up the status of man. Because animal life, as manifested in the senses and movement, was only a corollary of the stomach, animal life was degraded. When the digestive cavity is full, the animal sleeps. When it is empty, the animal goes out in search of food or prey. Hence neither the industry nor the cleverness of animals can be interpreted as signs of the existence of a soul. Only man has a soul, and this exalts him, in that he has not only bodily needs, which he shares with animals, but also needs of the soul, which are of another essence.

Proponents of the system of vegetation arrived at the same result by another route. Their spiritualist leanings were based on Hales's investigations of plant nutrition. Where did Buffon take his idea of the plant as machine, if not from his reading of *Vegetable Staticks?*[100] Hales had focused on the modalities of nutrition in both kingdoms. Clearly, then, it was hardly possible to differentiate between plants and animals on the basis of their ways of feeding themselves. The distinctive feature of the animal was not its digestive function but rather the fact that its structure was suitable for establishing relations with other animals. Still, the primary part of the animal's economy, comprising circulation and respiration, functions continuously and autonomously, and there was nothing to prevent reducing animal life to this: "With respect to its external functions, the animal seems to us almost identical to the vegetable; for, although the internal organization is different in plants and animals, the result is the same in both: they feed themselves, they grow, they develop, they will have the principles of an internal movement, they will possess a vegetable life." Take an animal, eliminate the principles of sensitivity and locomotion, and you have a plant. Conversely, take a plant, equip it with these two principles, and you have an animal: "Wrap these innards with an imaginary envelope, that is, provide them with senses and limbs, and animal life will not be long in making itself visible."[101] The animal, for Buffon, was no more than a roving plant, or, perhaps better, an awakened plant. Again, the status of the animal was diminished, in that the relations with other beings that are essential to its life are merely necessary consequences of its complex internal structure. Sensitivity and locomotion, in other words, are correlatives of need and not of a soul. At the same time, man was exalted. Although he might resemble animals in his material makeup, he differs from them in that he alone has internal sensations and therefore a soul. "This is more than is necessary to demonstrate the excellence of our nature, and the immense distance that the goodness of the creator has placed between man and beast; man is a reasonable being, animals are beings without reason."[102] Like Boerhaave, Buffon ultimately rallied to the side of the spiritualists, who recognized the existence of two substances, one immaterial and immortal, the other extended and mortal. This was the point that materialism, a system according to which there exists no soul separable from matter, and hence neither spirit nor God, sought to refute.

Materialism had two constituent parts. One is associated with the name of La Mettrie, who availed himself of the possibility of drawing irreligious consequences from the system proposed by Boerhaave. In order to conjure up that strange image, the "man-plant," what more had to be done than to substitute man for beast as the term of reference while keeping the plant in view as the other half of the comparison? Man and plants were said to be similar essentially with respect to the functions of circulation and respiration: "The lungs are our leaves; they take the place of this organ in plants.... Does a heart beat in every animal? Heat, nature's other heart, also causes juices to circulate in the tubes of plants, which breathe as we do.... In our species, as in vegetables, there is a main root and capillary roots. The reservoir in the loins and the thoracic canal constitute the former, and the lacteals the latter."[103] In La Mettrie's view—and on this point he agreed with Boerhaave—it was the absence of a digestive cavity that distinguished plants. Boerhaave and La Mettrie differed on the following point, however: Boerhaave held that the existence of a stomach was the criterion of animality, while La Mettrie held that it was the criterion of humanity. The status of man was consequently diminished, since he was merely a plant turned inside out. It follows that the faculty of thinking, like sensitivity and locomotion, has an anatomical substrate and is merely the result of the organization of the brain. Although the difference between man and animals was not thereby effaced, its status was changed from a difference of nature to one of degree: "The more needs an organized body has, the more means nature has provided for satisfying them. These means are the various degrees of that sagacity known as instinct in animals and as soul in man."[104] And that is not all. To deny the existence of a spiritual substance independent of matter was also to oppose idealism. The man-plant was therefore the metaphorical expression of a philosophical thesis. This no doubt accounts for La Mettrie's opposition not only to Cartesian innatism but also to Platonic idealism. Man is a terrestrial and not a celestial plant. "Man is therefore not an inverted tree, whose root is his brain, because the root is nothing more than the result of the fact that the abdominal vessels that form earliest grow together." [105] Hence there is no such thing as a "brain-root" or "seed of knowledge." In other words, there is no such thing as spontaneous generation of knowledge (idealism is a mythical account of the origins of knowledge), but rather learning and therefore subordination of the spiritual to the material.

There was also another possibility, however, of which Diderot availed himself, namely, to make a materialist reading of Buffon. The procedure was the same, to apply to man what Buffon had said about animals. This comes to the same thing as including man in the class of animals. Accordingly we are told that the only difference between, on the one hand, the species situated on the scale of perfection between the most perfect plant and the stupidest animal, and, on the other hand, "the class of other animals, like us, is that they are sleeping and we are awake, that we are animals that feel and they are animals that do not feel." [106] To translate, we are plants awakened, animated. Man has been "vegetalized" and thereby downgraded. If man were stripped down to the part of his economy that works continuously and autonomously, to respiration and circulation, he would be no different from a plant. To label all living things machines, all that remains is to establish a correspondence between the man who is thinking and the man who is sleeping: "I know of nothing so mechanical as a man absorbed in deep meditation, unless it be a man sunk in deep sleep." [107] Once again the difference between man and plants is one of degree rather than kind. The faculty of thought is therefore a function of need.

NUTRITION

The historian of nature (Buffon) here grants the metaphysicians far more than they would dare ask him for. No matter how we think once our soul has rid itself of its outer covering and emerged from its chrysalis, invariably the contemptible shell in which it remains imprisoned for a time has a prodigious influence on the order of thoughts that constitute its being: and despite the sometimes annoying consequences of this influence, it nevertheless is a clear demonstration of the wisdom of providence, which makes use of this stimulus to remind us constantly of the need to preserve ourselves and our species.[108]

Like La Mettrie, then, Diderot opposed the spiritualistic system defended by Boerhaave and Buffon. There was complete disagreement, moreover, on the problem of plant life.

It is of course true that "until the end of the eighteenth century not life but only living things existed . . . and if life could be spoken of at all, it was only as a character—in the taxonomic sense of the word—in the universal distribution of beings."[109] Accordingly one could define life as Boerhaave did, by way of a relatively complex character, namely, the existence of a digestive cavity intended to convert nutrients into a mass that could then be distributed throughout the economy. If one identifies life and nutrition, however, to the extent that nutrition depends on an organ such as the stomach, life must be denied to plants. Note, in passing, how interesting this choice of criterion is, for it leads to the inclusion in the class of living things of creatures that do not exhibit the most commonly accepted sign of animality, namely, movement. "That which provides animals with nourishment is always inside their body: and nutrition takes place even in animals that are naturally fixed and attached to some other body."[110] Thus the Cartesian ordering was completed and made more precise. Man is a living and thinking machine, whereas animals are living machines, "A hydraulic body that lives."[111] Plants, on the other hand, are only machines and are not alive. "By plant we mean a hydraulic body, which contains various kinds of fluids in different vessels, and which is attached to another body by one of its external parts, through which it draws from the body to which it is attached the matter that nourishes it and makes it grow."[112] One could also define life as Hales or Buffon did, however, using a much more general character, namely, the existence of a nutritive function, an operation by which a being draws nutritive matter from its alimentary environment, prepares that matter, and transforms it into its own substance. If life is then identified with nutrition defined in this way—thus neglecting the modalities of that nutrition—there is no reason to exclude plants from the class of living things. The dividing line now passes between the mineral kingdom and the other kingdoms. Even more, there are now "only two kingdoms in nature, one of which enjoys life, while the other is deprived of it."[113]

The choice of the criterion of life was a matter of metaphysics. To take one example, Hoffmann recognized, along with Hales and Buffon, that plant nutrition is accomplished by a relatively simple mechanism. This did not prevent him from denying, in the name of mechanism, that plants are alive. Was it not enough to adopt Descartes' criterion? "Although an advancing motion of the sap does occur in plants, it cannot be said that they are alive, because they have no heart and no warm, red blood. This explains why one does not say, properly speaking, that vegetables are alive or that they die, but rather that they exist or wilt."[114] Conversely the criterion of the living state proposed by Boerhaave hardly needs to be contested by anyone in favor of dynamism, because it is actually quite useful. Since it is hard to see how there can be life where there is no digestive cavity, it is enough to equip plants with a stomach if one wants to maintain, along with Kant, that they are alive. "Perhaps someone else might toy with the same concepts and say without laying himself open to reproach that 'the plant is an animal which has its stomach in the root (toward the outside).' ... Thus the plant possesses the principle of an internal life, which is vegetation."[115] In short, mechanism and dynamism govern the choice of criteria as well as their interpretation. Regardless of whether one supported the idea that plants are alive, as did Buffon and Kant, or opposed it, as did Boerhaave and Hoffmann, one's position was stated in the name of the same principle, namely, that life is not a metaphysical degree of being, but a property of matter. It was this point that animism, whose aim was obviously not to determine the specific nature of the phenomenon of life, sought to refute.

The solution suggested by the animists was not a novel one. What more were they doing than reviving the sentiments of Theophrastus, Pliny, and Columella concerning the soul of plants? "This soul acts in all living creatures, that is to say, in plants as well as animals, because it seems as absurd to me to try to explain the vegetation of plants by mechanical laws alone as it is to explain the economy of animals by those laws."[116] Vegetation therefore requires the existence of a principle responsible for executing the functions assigned to it: "Undoubtedly something more is going on here than a simple meeting of large and small pores designed in one way or another; indeed the source of the determination that occurs in trees must be sought farther afield, and it must be admitted that the principle of life that we have said animates them is a necessary, compelled agent."[117] In short, the only question that can be asked is this: Where does this principle or agent reside? One possibility is to accept the opinion of the ancients and locate the vegetating soul in the pith, more precisely at the junction between stem and root: "The principle of life seems to me to reside solely between the stem that rises and the root that descends; one may well cut off the head or shorten the roots of the plant—as long as no harm comes to the place where this principle of life has established its seat, the tree will not become any less vigorous as a result."[118] Of course this localization of the principle holds good only for trees, because "in other plants it resides exclusively in the bulbs or onion, as in tulips, hyacinths, imperials, anemones, etc."[119] There is nothing, finally, to prevent distributing this soul throughout the vegetable. Since plants also grow from cuttings and runners, it was necessary to envisage the existence of a soul that is not unitary but equally distributed throughout the parts of the plant and therefore divisible.

At last, the source of contention between the mechanists and vegetationists becomes clear. To prove the existence of mechanism in plants, the former chose quite specific and well-characterized functions (such as respiration and circulation) as models. An effort was made to follow the examples offered by animal physiology. Investigators were of course led to employ dubious and highly unreliable identification techniques, since the object, from the first, was to interpret the properties of plants. In other words, the object was to detect structural and functional similarities in a systematic manner, even though nothing guaranteed that the search for such similarities was well founded. By contrast, the supporters of vegetation tried to establish the existence of the system by reference to an object one step removed from the functions, the quantities of food absorbed, transported, and perspired by man. This brought them quickly to the crux of the matter. Either the economy of vegetables was similar to that of man or it was different. To decide which, the investigator was forced to compare the quantities of food absorbed, transported, and perspired by plants and men. This comparison revealed differences, which indicated that the two economies were indeed different. Here, animal physiology served as a term of comparison. What was the essential step in the articulation of the system of vegetation, if not to translate quantitative differences into qualitative terms?

Furthermore, when the circulationists carried out experiments, they invariably succeeded in confirming whatever theory they held to be true. By contrast, for Hales, the system of vegetation was merely a hypothesis, which he sought to verify through experiment. The mechanism of the plant was subject to constant modification, to disassembly followed by reassembly on a new pattern, so that the mechanists' image of vegetable life was constantly fluctuating and was forced to carry more weight than it could bear. In contrast, the system of vegetation was both simpler, since one had only to measure what entered and left the plant, and more consistent, because rather than call the initial construction into question, the results were steadily refined.

The mechanists, in establishing their theories of circulation and respiration, recognized similarities between plants and animals in regard to these most essential functions. This led them to make digestion the criterion of animality. By contrast, the vegetationists, in establishing the alternating motion of the sap, took note of structural dissimilarities between plants and animals. At the same time, however, they emphasized the similarity in the nutritive systems. Hence the difference between plants and animals could be based only on the way they feed themselves. Since animal life is determined by the complexity of the nutritive apparatus responsible for maintaining vegetative life in the animal, to which there corresponds an analogous apparatus in plants, the difference between the two classes of organized beings must lie in their relations to other creatures and not in the existence of a digestive cavity.

Finally, spiritualism and materialism were themes that could always be related to the concepts of plant physiology. Opposing concepts could be used to support the same metaphysical theme. Indeed, there is a direct path from the theory of the circulation of the sap to the definition of the animal proposed by Boerhaave (viz., An animal is a plant turned inside out). A similar remark holds good for the system of vegetation and the definition of the animal proposed by Buffon (The animal is an awakened plant). Although these definitions bring different concepts into play, they have exactly the same function; by defining animals in terms of plants, Boerhaave and Buffon were downgrading the status of the animal and at the same time exalting the status of man (spiritualism). Thus one of the metaphysical implications of Cartesianism was rediscovered. The same concepts, however, can be used to support different themes. Indeed, it was by using the concepts put forward by Boerhaave that La Mettrie developed a materialistic philosophy. The same observation can be made with respect to the concept used by Boerhaave and Kant. On the one hand, the two men agreed in defining life by the existence of a digestive cavity. But they clashed over the question of whether plants are alive, Boerhaave denying that they are (mechanism), and Kant taking the opposite view (vitalism).

3 Generation

At the end of the seventeenth century, a surprising divergence emerged within preformationism. For the preformationist thesis applied to both plants and animals, but different mechanisms were thought to activate the germ in each. In animals, the first nutrient came from one of the two parents, and the delicacy of the nutriment was in proportion to the delicacy of the embryo. In plants, on the other hand, the germ within the seed received its first food from the earth. Surely the crude aliment in the earth could not possibly be appropriate for the plantlet, whose structure was just as delicate as that of the animal embryo. It was therefore necessary to interpolate a further mechanism, analogous to the animal mechanism, between the preformation stage and the nutrition that made it possible for the plantlet to survive and grow. In other words, what was needed was a generation function, a "dynamo of growth." [1]

In order to investigate this mechanism in plants, an analogy had to be found in the animal kingdom that could explain the workings of a function as essential as nutrition was presumed to be. A problem cropped up immediately, however: animals reproduce in a variety of ways. Insects do not reproduce in the same way as gallinaceans, and gallinaceans do not reproduce in the same way as mammals or fish. Thus the choice of model turns out to be a delicate problem. If it is not to be arbitrary, the choice of model should be the result of a direct comparison between the seed and some animal product. There are only two ways to do this. Either the seed can be compared to the egg, which leads to the assumption of sexuality in plants. Or the seed may be compared to the offspring of mammals, in which case some process similar to that by which the fetus is fed must be imagined. I shall call those who choose the first alternative *sexualists*, and those who choose the second *agamists*.[2]

In histories of botany, it is common to draw as sharp a contrast between sexualists and agamists as between the proponents of analogy and the champions of observation and experimentation. Some botanists identified the seed with the fetus and viewed the flower as an excretory organ. Ovists favored this system throughout the eighteenth century. Others took an experimental approach. Tournefort, Spallanzani, and Alston went far afield, we are told, because they relied on analogies. Not only are they supposed to have failed to notice the similarity between generation in plants and generation in animals, but, as puritans, they presumably could not help denying the existence of a kind of sexuality in plants. By contrast, Camerarius, Bradley, and Linnaeus are said to have demonstrated the existence of this essential function through experiments with the organs of flowers. Not only are they supposed to have shown how reproduction in plants resembles reproduction in animals, but we are told that as unprejudiced scientists they saw clearly in plants what in man is concealed from view. The heart of the matter lies not in these conflicts but in the alternative that made it possible to resolve the same problem in two different ways. Put differently, the point is to consider the mechanism that activates the preformed germ in plants.

The comparison of the seed with the egg was not a novel one. Empedocles long ago believed that the seed of a plant was analogous to the germ of an animal embryo. But the sexualists developed this analogy to the full, using a procedure that in essence amounted to choosing oviparous animals as the term of comparison. Since the seed resembles the egg, like the egg it may be either fertile or sterile. Now, just as incubation reveals the nature of the egg—if it has been fertilized, it produces a bird, and if it has not been fertilized, it rots—so sowing makes it possible to determine the nature of the seed. Touching it can furnish additional indications. A hard, full seed is usually likely to germinate, whereas a limp, empty seed will dry out rapidly. Thus examination of the seed, which may be either fertile or sterile, makes it possible to assert the existence of plant sexuality in no uncertain terms. "I would not deny that sometimes trees and plants bear fruit and bring them to maturity without the intervention of a male or dispersal of semen. For even though most birds do not lay eggs unless there has been some relation with a male, nevertheless, certain birds, such as hens, do lay without prior coitus rather frequently, although the eggs they lay are germless and infertile."[3] The comparison with birds and hens leads to the idea that the male plant has the power to fertilize the seed, and that the female plant can produce seed that is either fertile or sterile. What we have here is something quite different from the old distinction between males and females, for the terms had quite a different sense in the then current usage. Males were characterized by the absence of seed and females by the presence.

If one searches for a relationship between the sexes in plants similar to that found in animals, a problem crops up at once. Plants do not manifest their masculinity or femininity in differences of size and shape. There is an easy way to get around this difficulty, however: merely identify which parts need to be examined in order to differentiate unambiguously between males and females. In other words, attention must be focused on the external parts of plants, and their contents must be examined. The structure of the container, and especially of the contents, is apt to reveal the nature of the organ, because the egg sits inside the ovary of the hen, whereas the sperm issues from the penis. Now, in plants, it was observed that the seeds sit inside the pod or husk, and the pollen originates from the tips of the anthers. "Hence it seems rational to denote these apices by a more noble name and attribute to them the importance of masculine sexual organs; it is there that the semen, the powder that constitutes the subtlest part of the plant, accumulates, and it is from there that it later flows forth. It goes without saying that the ovary and the style that is part of it are the female sexual organs of the plant."[4]

The study of flowers came to be viewed as related to the sexual function. It is true that Renaissance botanists had already given names to the various parts surrounding the petals. This work was only descriptive, however. To ascribe meaning to the terms of this nomenclature, all that was necessary was for the parts of flowers to become organs associated with a function. Most of the terminology was borrowed from the vocabulary of the arts and based on similarity of form. The female organ was denoted by a term of Greek origin: *pistil*, which means column. A column is composed of three parts: the base, the shank, and the capital; similarly, the pistil consists of a base, which contains the seed, a style, and a stigma. The male organ was called the *stamen*, a word of Latin origin denoting the thread in the warp of a loom. It is made up of two parts, the filament and the anther; the pollen exits through the dehiscence of the latter. In keeping with the model chosen, however, sexuality was first studied in plants that carry the reproductive organs on separate stalks. Before male and female parts can be observed within a single corolla, the nature of their appearance must be understood. Thus, after examining the distribution of the sexes, Camerarius arrived at a classification into three groups. "The stamens and pistils are either united in one flower, or they grow separately on different branches of the same stalk, or, again, they grow on separate stalks."[5] Parallel to this sexual topography was a floral chronology. For the flower always precedes the fruit; the unfolding of the petals and the male parts is followed by their withering. At the same time, the lower part of the pistil grows larger and its upper portion wilts. The falling of the petals and the appearance of the fruit attest to what went before, namely, fertilization. The oft-repeated assertion that flowers must come before fruits and seeds was taken as a general rule.

The idea of basing plant reproduction on an analogy borrowed from the animal kingdom was a hypothesis. A series of experiments was devised in order to verify the existence of sexuality. To prevent fertilization of the seeds, steps were taken to prevent the stamens from communicating with the pistils. The latter had to be removed from the influence of the former. Doing this should assure that the seeds turn out sterile. Conversely the seeds should be fertile if nothing stands in the way of contact, proof by contradiction. In species where male and female stalks are separate, isolation of the female stalks neutralizes the effect of the pollen. What is more, this happens in nature. "A mulberry tree in whose neighborhood no other mulberry bearing flowers was to be found produced mulberries that contained no germ."[6] With species bearing staminiferous and pistiliferous flowers on the same stalk, the technique used was a crude one. The staminiferous flowers were simply pulled off. "Now so soon as the *Catkins* appear, they must be carefully taken from from the *Tree*, and it will produce no *Fruit* that Year."[7] In hermaphroditic flowers, resection of the stamens well before the fertilizing dust has ripened suffices to prevent the production of seed. For Bradley, the use of a control apparently made the experiment more cogent. "These *Tulips* being thus *castrated*, bare no *Seed* that Summer, while on the other hand, every one of the 400 *Plants* I had let alone produced *Seed*."[8] Gaston Bonnier maintained that it was the choice of the tulip that, unbeknownst to the experimenter, had made the operation successful: "Because he used tulips, Bradley was dealing with a flower that is an exception to the rule, in that it produces no sweet liquid and therefore is not visited by melliferous insects. Had he worked with almost any native flower, with borages, for example, he would have seen fruits and seeds produced by all the flowers whose stamens he had cut out. What would he have concluded then?"[9] This question is easy to answer, provided, of course, that one looks at the texts. At about the same time, in fact, Miller obtained fertile seeds even though he had carefully cut off the stamens of a dozen tulips. He therefore repeated the experiment, this time with controls. He noticed how the bees carried on: "Near one tulip plant, I saw some bees very busy amidst the flowers. I watched them fly away with their legs and stomachs laden with dust, and one of them landed on a tulip that I had castrated."[10] The bee, then, was seen as a cause of disruption of the experiment rather than as an agent for transporting pollen. Without a theory of pollination it was hardly possible to interpret this observation. The only question at issue at the beginning of the eighteenth century was the existence of sexuality. Stated differently, the problem was to show that the sexual organs are essential to the work of reproduction.

How generation is accomplished in plants remained to be established. Clearly the relationship between the stamens and the pistil was nothing like coitus, at least not in species in which male and female flowers are separate, whether on the same stalk or on separate stalks. In hermaphrodites, though, the possibility of sexual union was considered. Obviously this involved a comparison with copulation in animals. It cannot be overemphasized, however, that in this period a surprising difference was noted between the two kingdoms in regard to the distribution of the sexes. Thanks to the work of Swammerdam, who described the love life of snails, hermaphroditism in animals came to be accepted as natural. The androgyne, once thought to be a freak or caprice of nature, was no longer seen merely as a mythical figure. This made comparison of one kingdom with another possible and led to the identification of unisexual and bisexual species in plants and animals. Now, as Blair observed, "although the number of hermaphroditic animal species is small, hermaphrodites constitute the largest class of plants."[11] Thus, in the plant world, in contrast to the animal world, bisexuality was the rule and unisexuality the exception. This contrast is crucial, for it structured the representation of the sexual mechanisms in the different species of vegetables.

As for the rule, as opposed to the exception, it is immediately clear why the contrast between vegetable bisexuality and animal unisexuality justified identifying the sexual mechanics of hermaphrodites with copulation of a certain sort. In animals, in effect, unisexuality required mobile structures. By contrast, the bisexuality of most plants was related to their being fixed in one place. Hermaphroditism made up for the absence of a locomotive function. Thus the idea that self-fertilization results from the union of the stamens and the pistil came to the fore: "*Plants* wanting *local Motion*, require therefore this Union of *Sexes* in themselves, by which Means they may generate without the Neighbourhood of other *Plants*."[12] As a result, observers did not merely report the dissemination of pollen but generally added to their reports a description that could just as well have been of coitus in animals. This probably accounts for the attribution of so important a role to the corolla, which surrounds the sexual organs of plants: "Ordinarily, it is only after they have consummated their marriage that the corolla allows them to show themselves; or, if it opens slightly while they are locked in one another's embrace, it blossoms fully only after they have separated."[13] Another point weighed in favor of this view: hermaphroditism existed only in myth. Embodied in a creature of flesh and blood or fiber and sap, it was a monstrosity, because it evoked a twofold impotence. This led to personification of the sexual organs, or, in other words, to a substitution of the male-female couple (the corolla being the nuptial bed) for the neuter or asexual image evoked by the idea of bisexuality. Concomitantly, monstrosity was now perceived in a different place, in any case by Vaillant, who thought it perverse that the petals multiplied at the expense of the genital parts, symbolic of castration. "Upon those agreeable monsters, the double flowers, we lavish care; rarely if ever do we find them with testicles, which these unnatural mothers devour, from the cradle, as it were, by taking all the nourishment for themselves."[14]

The mechanics of sex in unisexual vegetables was obviously more problematic than the mechanics of sex in hermaphrodites. Botanists who justified the existence of bisexual plants by pointing out that plants are immobile creatures could not help being disconcerted by the existence of a small number of unisexual vegetables. There was a very simple way around the problem, however. Since unisexual plants are exceptions to the rule, why not explain the mechanics of their sexuality by using a natural, but likewise exceptional, expedient? Thus insects were assigned the responsibility of transporting the fertilizing material. Conceived in this way, the mechanistic theory of pollination by insects is far from being a precursor of a theory of pollination, being merely a response to a requirement of logic. "Plants may be impregnated by Insects much smaller than Bees; for as the Creator of all Things, in his infinite Wisdom, appointed this Way of Generation to Vegetables, which are incapable of Motion to each other, it may be supposed, that he had so ordained it, that a small Part of the Male Dust should be sufficient to perform that Office."[15] This sort of observation was generally combined with travelers' descriptions of the fertilization of palm trees. Descriptions of this practice showed plainly that man served as intermediary. Still, generation in unisexual plants remained ambiguous, on the borderline between the real and the fabulous. Most botanists evoked the loves of two palms planted fifteen miles apart, one at Otranto, the other at Brindisi. The model most certainly involved sublimation, for the wind was seen as love's messenger, and, again, the plants were personified.

For reasons of anthropomorphic prejudice, sexualists placed plants above animals. But identifying inferior beings with superior ones was incompatible with the idea of a hierarchy of living forms. For this reason, the opposite approach was usually taken. Comparisons were still made between plants and animals, but the animals were chosen from a different portion of the animal kingdom. By using animals of the lower classes, one could avoid the impasse to which anthropomorphism led, to say nothing of the inconsistency. It was possible, further, to prove by analogy that it was correct to compare the sexual mechanics of plants to that of animals. Rather than personify the sexes of plants, one could do as Bradley did and argue in favor of "this Union of *Sexes* in themselves, by which Means [plants] may generate without the Neighbourhood of other *Plants*; they are in this respect like *Mussels*, or other immovable Shell-fish, who are *Hermaphrodites* of this kind, having their Propagation without the help of one of the same Species."[16] With unisexual vegetables, exactly the same procedure was used. Instead of projecting what in man is an amorous aspiration of mythical dimension, it makes more sense to ask whether it is "more difficult to suppose that the Air is an Intermitter in the Fecundation of *Plants*, than to make the Element of Water an Intermitter in the Propagation of *Fish*, and chiefly of *Oysters* and other *Animals*, which lie *immovable* at the Bottom of the Sea." [17] Diminishing the status of plants serves to reclassify them in such a way that the hierarchy of living forms is respected.

Whether sexuality in animals serves as a model or as confirmation by analogy, it is clear that a distinction is being made between autogamy in hermaphrodites and cross-fertilization in unisexual plants. Thanks to this distinction, which involves the modalities of generation, the problem of species conservation is resolved for bisexual vegetables but not for unisexual ones. In the bisexuals the proximity of stamens and pistil not only makes up for the lack of a locomotive capacity but also guarantees the permanence of species. The conception of plant sexuality current at this time allowed an idea implicit in the myth of the androgyne to come to the surface, the idea of a creature engendering itself. What was hermaphroditism in plants, if not a living example of the fable? "Nature ordinarily seals within a single flower all the parts that must contribute to the conservation of the species." [18] The propagation of unisexual plants is exposed to various dangers, however, and nothing could be less certain. This idea of riskiness is reinforced by examination of the "intermitters," other than man, of course. The motion of the air, an inanimate agent, is eminently fortuitous, and the same holds true for the motion of insects, animate but unconscious agents, for the insect will fasten itself promiscuously to the flowers of one species after another, more a raider in search of booty than a mediator in the propagation of the species. In short, hermaphrodites lock themselves up to procreate, whereas unisexual plants need an agent that curiously seems to foil the intentions of the creator.

As a result, two distinct avenues of inquiry were open to sexualists concerned above all else to prove the existence of the sexual system. Since hermaphrodites fertilize themselves, research focused on identifying the organic links whose purpose is to facilitate the process of impregnation. The relative sizes of the sexual organs actually seem to be calculated so as to ensure that the flower's pollen will act on its stigma. Sometimes, however, the stamens are smaller than the pistils. But this occurs mainly in flowers that are bent over, so that their pollen easily reaches its destination. Most botanists were content to do as Bradley did, emphasizing the arrangement of the sexual parts. In flowers in which both sexes are united, "the *Pistillum* is always so placed that the *Apices* which surround it are either equal in Height with it, or above it, so that their *Dust* falls naturally upon it."[19] This interpretation of the facts is obviously based on the assumption that the reproductive organs grow at the same rate and reach maturity at the same time, for which confirmation was soon forthcoming. By contrast, the distance between male and female plants, as well as the unreliability of the intermediaries, did not facilitate, but rather complicated, the process of impregnation. Somewhat paradoxically, however, this provided the occasion for an argument in support, if not of regular propagation of the species, then at least of the sexual theory. For the existence of hybrids was an irrefutable proof of the union of male and female seeds. In this way it was possible to explain, as Dudley did, the mixture of colors in Indian corn: "The stamina, or principles of this wonderful copulation, or mixing of colours, are carried through the air by the wind."[20] The violation of the rule of endogamy only provided unambiguous proof that sexual reproduction is indeed the rule. Bisexuality was therefore an assurance of legitimate marriages, while unisexuality seemed to favor illegitimate marriages. This was of far less importance for the time being, however, than the proofs of the theory of plant sexuality deduced from the existence of illegitimacy.

As for the internal mechanism of fertilization, the final

touch to the system, the approach taken in plants was the opposite of that by which the question had been resolved in animals. Examination under the microscope of sperm and ovaries had led ovists to locate the germ in the female and animalculists to locate it in the male. In plants the location assigned to the generation process depended on the initial choice of theory. Because plants have a variety of sexual structures, it was quite easy to fit the facts to one doctrine or the other. Examples abound, some favorable to ovism, others favorable to animalculism. One possibility was to accept Leeuwenhoek's theory, as Claude Geoffroy did. This meant that "the dusts of the flowers are the first germs of plants, which in order to grow need juice that they find in the embryos of the seeds, just as animals need the egg and the uterus before they can see the light of day."[21] The stigma, which bristles with fine down and is usually coated with a viscous fluid, is thus a receptor organ, its role to receive, grasp, and retain the pollen. The style, which is the conductor organ, must be pierced with tiny holes like the spout of a watering can, since the powder has to pass through it in order to reach the egg. There was no problem about how the germ enters the seed, for observation revealed a tiny opening in the seed quite similar to the opening in animal eggs known as the cicatricle. Thus there was a plainly visible orifice that led directly to the plantlet, through which the seed could be entered. The other option was to adopt the ovist theory, as Vaillant did. "Since nature always acts according to uniform laws, it must be concluded that what takes place on this occasion in animals takes place in the same way in vegetables.... The vapor, or volatile spirit, that is given off by the grains of dust, proceeds to fertilize the eggs."[22] The stigma is still the receptor organ that receives the pollen, but its surface has no perforations and the style is solid. This is because the puff of vapor that has to be conveyed to the germ in the egg follows a path appropriate to its nature, flowing along the tracheae that traverse the pistil and lead to the surface of the stigma. The vapor thus passes through the tracheae into the seed via the umbilical cord, carrying the first nutriment to the seedling.

Until around 1770, the study of generation revolved around the three questions whose outlines I have just sketched as they appeared at the beginning of the eighteenth century: the existence of sexuality, the sexual mechanism, and the location of the plantlet. Plant physiology therefore remained in thrall to animal physiology. Nevertheless, this dependence did not prevent botanists from making a number of contributions and corrections to the theories they borrowed whose import cannot be neglected. In the first place, further experimentation was possible. For one thing, experiments could be done to show beyond a shadow of a doubt that the stamen and pistil are indeed the organs responsible for conception. For another, there was reason to investigate the role of the pollen, because "the surmise concerning the pollen which has been remarked, as that which might correspond to the masculine semen, [is] not yet very clear."[23] The first series of experiments was distinguished from earlier work by the fact that the conditions of observation had been improved markedly. Rather than work in open fields, experimenters isolated the objects of study in greenhouses or closed rooms. Potted plants that could be moved easily and thus kept isolated were used. This procedure made it possible to eliminate extraneous factors that might have disturbed the experiments. Thus the possible effects of insects and of the wind were forestalled, enhancing the credibility of the results. Experimentation with potted plants isolated in greenhouses offered another advantage. It became possible to explain, for example, why Camerarius had failed in one case to obtain the expected result. Had he not reported that, despite the absence of male stalks, fertile seeds had appeared on female stalks of hemp? Linnaeus was not disconcerted by this result, which he regarded as due to a defect in the experimental procedure: "I was thereby completely convinced that the examples reported by various authors of female stalks of hemp that produced good seed without the aid of males were not correct, because the females had been made fertile by the seminal dust that the wind had brought to their stigmata from afar."[24] Another advantage was that the experimental conditions could be varied. In hermaphroditic flowers the style was removed rather than the male organs. When the style was cut before it received the dust from the anthers, the fruit died. This was sufficient proof that the style is indeed the female organ. The interest of this type of experiment is immediately apparent, especially if the object of study is chosen judiciously, as in Logan's work, which was focused on grains. All one had to do was cut off a certain number of styles and then observe that the corresponding locules became sterile. "In the ears of corn from which I had removed several styles, I found as many ripe seeds as I had left styles."[25]

The second series of experiments, unprecedented in nature, focused directly on pollen as the object of study. One procedure was to attempt artificial fertilization, and another was to prevent the powder from coming into contact with the stigma. The point was not to determine how the powder acts but rather to show that it acts alone, as the sole agent of fertilization. In other words the object of the experiments was to show that the seed becomes fertile if and only if the female organ comes into contact with the pollen. One way of doing this was to take dust from the anthers and throw it by hand onto the flowers of the female palm, as Gleditsch did. Logan, on the other hand, focused exclusively on motion of the air and its effects on the pollen. By using a piece of fine muslin cloth to isolate the female panicle of a corn plant shortly before the styles appeared, he was able to prevent the dust from coming into contact with the stigmata. These experiments proved conclusive.

Gleditsch planted the seeds he obtained and observed that "they gave birth to plants in conformity with their origins, that is, to small palms, which shows incontrovertibly that fertilization was fully accomplished."[26] And Logan found that "in the ears covered with muslin, not a single ripe grain was visible; the seeds were no more than empty shells."[27]

The study of sexual mechanics also progressed. Because the sexual organs are not so conveniently arranged in all creatures, hermaphrodites were investigated first; the unisexual species were studied because of the unreliability of insects and winds as agents of reproduction, opening up the possibility of creating hybrids. Gleichen observed that because these agents "have little knowledge of the mysteries of Botany, they might well carry the dust of a lowly flower such as that of the squash onto a precious flower such as that of a melon, an acorn, a walnut, etc."[28] Thus attention had to be paid not only to the way in which nature neutralizes her agents' effects but also to how she accommodates herself to them, since the propagation of unisexual species depends on their assistance. The sexual mechanism of hermaphrodites was apparently less reliable in certain species than in others. Indeed, in some species the arrangement of the reproductive parts tends to thwart autogamy rather than ensure its success. The primulus, the bellwort, and the iris have such peculiar structures! There are no organic links to facilitate the action of the pollen on the stigma located in the same flower as the anther from which it was released. The odd construction of these flowers was not, however, seen as an obstacle to self-fertilization but rather as an opportunity for the sexualists to exercise their ingenuity. Their problem was to figure out reliable ways in which self-pollination might be accomplished in these flowers in spite of the obstacles. To take one example, in flowers where the pistil is longer than the stamens, "the creator has caused these flowers to lean over so that the pollen can fall more easily on the stigma, as in the *primulus* and *bellwort*,

Frederick William Gleichen, p. 39 from *Découvertes les plus nouvelles dans le règne végétal* (Nuremberg, 1770). Photo courtesy of the Muséum national d'histoire naturelle, Paris.

GENERATION

etc."[29] Adanson added a further explanation: "If the stigma is longer than the stamens, then it bends toward the anthers, resuming its erect posture only after fertilization, as in chickweed, granadilla, and corn poppy. Or else it grows only when the stamens are mature, and it covers itself with dust as it grows past them, as in the Compositae and some bellworts."[30] Using similar arguments, botanists eventually succeeded in explaining away all peculiarities, which were subsumed under the general rule of autogamy.

The sexual mechanism of unisexual flowers presented a more embarrassing problem, because it bore on the question of the conservation of species. Apparently only two solutions were possible: either the agent of fertilization is unreliable and hybrids are the result of its doings, or the agent is reliable, which would explain the constancy of species but leave the existence of hybrids as a puzzling anomaly. Actually the problem that confronted botanists was similar to one that had come up in the study of animals; the theoretical status of hybrid plants was the same as that of hybrid animals. "In view of the harmony that reigns everywhere in the universe, every reasonable philosopher must first of all believe that anomalies such as these are governed by laws and have limits of their own."[31] Since nature acts in both kingdoms in the same way, regular propagation must be the predominant kind; this did not, however, rule out the possibility of cross-breeding, if not among species in general, then at least among closely related species. If species persist, then it was assumed the agents of fertilization must be infallible. Though insects and winds appeared unreliable, they must in fact be perfectly adapted to their mission. At this point in the argument an analogy was introduced, the artificial fertilization of date palms. Man, the conveyor of the pollen in this case, is conscious of what he must do in order to obtain the dates he is after. Gleditsch accordingly argued that, even though the motion of insects and winds is natural, it is merely "another artificial means of fertilization, which takes place in the open air and cannot be obstructed by preventive measures. . . . Bees and many other insects alight on flowers and have bodies that are partially covered with hair; they accomplish the work of artificial fertilization without other assistance, and are the sole means for doing so."[32] The artificer here, in contrast to the case of the date palms, is not the agent of fertilization but rather nature herself. The argument harks back to the old idea of a natural economy, rounding it off. Nature is wise, farsighted, and cunning. She sees to it not only that the seed will be disseminated but also that the pollen will be carried where it needs to go. From sowing to fertilization, nature takes a hand at every step along the way. In short, the transport of seed and pollen is part of the natural order of things. Fertilization is accomplished "as nature sees fit, by the motion of the air, by insects, or by human industry."[33]

To prove that insects and winds are reliable agents of fertilization, however, it is not enough merely to invoke nature's intentions. How does nature make use of the haphazard behavior of insects, and how does she take advantage of the uncertain winds? These questions, too, must be answered. One problem crops up immediately: How does nature nullify the effects of her intermediaries when she has to? The sexualists had a very simple answer to this question. Assuming that there are specific affinities between the dusts and germs of each species, not all dusts are compatible with all germs. The dust of a given species will result in regular propagation when it encounters a germ of the same species and in a hybrid when it encounters the germ of a closely related species. "There is no reason for surprise, then, that from the scabious and the campion, say, or from the horehound and melissa . . . and other species whose flowers nature has made to show great affinity of structure, there should now and then be born new plants or hybrids never before seen."[34] But a given dust will also be incompatible with some germs, from which it follows that species do not mix: "Just as I do not believe that the seminal fluid of a rabbit can cause the germ of a chicken to develop fully, so I do not think that the dusts of the lily can fertilize the seeds of a pear tree."[35] Mere neutralization is not enough, however, because compatibility wins out over incompatibility, as the regularity of propagation in unisexual species attests. This raises a second problem. How does nature turn the situation around so as to favor propagation? How does she make up for the small probability of action? An easy way to resolve the problem is to assume that both the quantity of fecundating matter and the number of females are large. For one thing, as Gleichen observes, "everything here is divine, everything reveals the creator. What an astonishing quantity of grains of dust God causes to be produced and formed by nature's most secret springs!"[36] For another, "providence has seen to it that, in plants with separate individuals, there are more females than males."[37]

Finally, in order to overcome the discrepancies between ovism and animalculism, more careful observations were needed. Both seed and pollen were therefore subjected to microscopic examination. In the debate over whether the leading role was to be assigned to the seed or the pollen, pollen ultimately won out when globules were discovered. Needham was investigating the dust from the anthers. He first placed the pollen under the lens of the instrument and then added a small drop of water to the slide: "In the space of a few seconds, I saw a distinct trail of globules enveloped in a membranous substance, which darted out from the grains of dust."[38] These were thought to be the first principles of the plant's existence, more or less the equivalent of the animalcules in the semen. This discovery made it clear that the internal mechanism of generation would have to be reconsidered, since not the pollen itself but only its contents enters the seed. Thus the problem was not to revise the description of the pistil but to rework it completely, from stigma to ovary. This task was made easier by the fact that the circumstances surrounding the observation suggested a new model. What caused the grains of pollen to burst and set the globules free, if not the water? Exactly the same thing occurs when the pollen falls on the stigma, which is always moist. "The force of the action of these grains is what throws the fertilizing substance into the ducts of the pistil that lead to the ovary."[39] Thus the stigma is a receptor organ that also plays a part in a propulsive mechanism. After fertilization, a greenish spot can be observed inside the seed, and this was identified with the globule, which contains the plantlet and is originally part of the shell of the pollen grain. As François Jacob has observed, however, "throughout the eighteenth century, and as long as living things were seen as combinations of visible elements, preformation and preexistence offered the only possible solution to the problem of generation."[40] Thus, in this period sexuality was merely an object of observation, an object for the taxonomist. The only possibility was to shed some light on the reproductive apparatus, the visible instrument that assures the perpetuation of the species. Sexualists therefore carried on by trying to generalize the system.

Nature, it was generally held, conducts her operations with apparent uniformity. The smaller a living thing is, the more hidden its attributes. In the so-called cryptogams, plants in which the reproductive apparatus is difficult to see, sexualist theories had to be extended systematically. But it was impossible to be sure that the reproductive organs in the cryptogams were copies of those in the phanerogams, since the presumed structural similarities could not be detected; thus it was impossible to be sure that no plant is without sexuality. "Marriages" were observable only in phanerogams, so the necessary evidence was lacking. This being the case, the concept of sexuality could be extended to cryptogams only with the aid of microscopic scrutiny of plant forms. The existence of the organs was not doubted;

the problem lay rather with their small size. Microscope and magnifying glass were certainly useful tools for exhibiting the sexual apparatus of cryptogams, but the way they were used depended on the model that guided research, namely, the sexuality of the phanerogam, which was presumed to define the nature of sexuality even in plants whose sexual organs could not be seen. Reducing sexuality to the bare essentials, a fecundating fluid and a seed, the sexualists were able to suggest a plentiful variety of models. A technique could then be chosen to identify the "sexual parts" in the cryptogams; since what was not looked for was not seen, and what was not seen was regarded as nonexistent, the technique always proved adequate to the task. There was yet another advantage to this manner of proceeding: observations of animal life could be adduced as confirmation of theories about plants. "The organs of generation are not the same in all animals, and, similarly, the organs of fructification are not the same in all vegetables." In passing, it should be noted that alternating generations in mosses and ferns could not have been suspected at this time for a very simple reason: "Although the extreme fragility of most mushrooms and other cryptogamic vegetables is quite in keeping with nature's plan, it is a stubborn obstacle to the study of germination in these plants."[41]

For ferns, which are similar in appearance to phanerogams, a relatively ponderous and complex model was proposed. The sexualist had a choice of emphasizing the structure of either the "containers" (stamen, pistil) or the "contents" (pollen, seed). If one followed Micheli and chose the first option, attention was focused on the organ's structure as an index of its nature. The stamen, in other words, differs in form from the pistil. Now, in ferns, what do we find in the neighborhood of the seeds but a nipple surmounted by hair? These parts of the fern obviously correspond to the stamens with their anthers. If, on the other hand, one followed Bernard de Jussieu and chose the second option, attention was focused on the structure of the contents rather than the container. Seed and pollen differ not only in form but also in size. What do we find in the "buds" of ferns but locules containing both small, round "seeds" and larger, oblong "seeds"? The former were identified with the grains of pollen and said to be contained in anthers belonging to stamens without filaments. The latter were identified with the seeds in a pistil without style or stigma. Thus, in the pillwort "each quarter of the globule is filled by a hermaphroditic flower composed of stamens and pistils, arranged on a common placenta."[42] Réaumur studied both contents and containers in the fucus and consequently identified the filaments as stamens without anthers and certain encapsulated grains as seeds.[43]

The sexual parts of mosses and fungi were harder to see, however. Corpuscles exhibited no differences in shape or size. Thus a problem arose as soon as one tried to designate particular parts as seed or pollen. Fortunately there was a very simple way to get around this problem, namely, by making use of whatever properties the corpuscles might exhibit. These properties were actually indications of the nature of the corpuscle. For the seed, the relevant property was the faculty of germination; for the pollen, it was the mode of dissemination, namely, a sudden explosion. Since both properties could be found in the same corpuscle, however, two opposing systems could be proposed. If one followed Hedwig and chose germination, then the parts found in the capsule had to be identified with seeds. The other corpuscles were identified with the pollen in the anther.[44] If one followed Linnaeus and chose the mode of dissemination, however, then one arrived at a diametrically opposite result, namely, that the corpuscles in the capsule correspond to pollen. Conversely what Hedwig thought was pollen Linnaeus regarded as seed. Fungi posed a similar problem. The difficulty was to distinguish the seed from the pollen, to give a precise description of the essential difference between the seeds of mushrooms and the globules of the fertilizing dust, whatever it might be. Rather than use the distinctive features singled out by Linnaeus or Hedwig, Bulliard turned to an investigation of the content of the corpuscles. In phanerogams, when the pollen grain is immersed in water, the shell explodes, liberating a viscous fluid; whereas under the same conditions the seed inflates and splits but does not expel any fluid. By subjecting various corpuscles to heat, Bulliard observed that in some cases the grains burst, but in others there was no change. The former were identified with pollen, the latter with seeds. "In mushrooms that have sperm ducts it is exactly the same. The germs are preexistent. At a certain time they are penetrated by a fluid similar to that found in the globules of a fertilizing dust.... Hence there is a perfect analogy between the fertilization of the seeds of these mushrooms and the fertilization of vegetables that have powdery stamens."[45]

If the sexuality of cryptogams is seen in terms of the sexuality of phanerogams, problems of two kinds arise. First, fertilization occurs in different circumstances. Cryptogams are usually found in an aquatic environment rather than in the open air. Second, the position of the flowers is puzzling. In higher plants the sexual apparatus is always visible. This is not the case with cryptogams. The disparity between phanerogams and cryptogams as to fertilization can be reduced in two ways. One can first of all point to the exceptions. There are some vegetables, like the fig tree, whose flowers "are hidden beneath coverings. Once only a single example of this type was known, but now the flower of the pillwort provides another."[46] The other way to eliminate the difficulty was to assume that cryptogams are fertilized under the same conditions as phanerogams. Thus it was possible to imagine, as Guettard did, that the marsilea is fertilized when the water recedes. "This assumption eliminates all the difficulties, and the fertilization of the marsilea is subsumed under the general law governing terrestrial plants."[47] The problem is more delicate with algae, however; since they are always under water, the pollen would quickly be diluted. To get around this difficulty, it is sufficient to assume that these plants are hermaphroditic rather than unisexual. For Correa de Serra "the conceptacles of all plants containing seeds and mucus must be considered to be hermaphroditic flowers; the seeds they contain are like ordinary seeds, and the mucous substance is like pollen."[48]

AGAMISM

Another theory was put forward in the classical era in answer to the same question, the theory of agamism. Instead of taking generation in the oviparous animals as their model, as the sexualists did, agamists chose the viviparous animals. This may seem paradoxical. How could the agamists have arrived at a theory of asexual reproduction by comparing the seed to the offspring of mammals? The birth of a fully developed living creature not protected by a shell of any sort presupposes a prior act of fertilization. How, then, did the theory of agamic reproduction in plants result from importing a viviparous model from animal physiology into plant physiology? To begin, it should be noted that the preformationist thesis was based not so much on the structure of the animal embryo as on the structure of the seed. In plants, the seed exhibits a miniature version of the future plant in a quite visible way. We see the curled-up plantlet embedded between the cotyledons. The plant seed was thus a point of reference, indeed an argument in favor of preformation. This point was emphasized by Maupertuis: "Based on an analogy with plants, where what appears to be production of the parts is in reality merely development of parts already formed in the seed or bulb, most modern physicists, being unable to understand how an organized body might be produced, hope to reduce all generation to mere development. They believe it is simpler to assume that all the animals of each species were already contained, fully formed, in a single father or a single mother, than to allow for any new production."[49] This makes it clear why agamists were eager to establish a reciprocal relationship between the seed and the fully formed young animal. Such a relationship would make it possible to understand the mechanism of generation in plants. This point is crucial: the seed, which is visible, can be compared only with the fetus or possibly the fertilized egg.

The comparison brings two processes to light, the one not an exact replica of the other: in animals the germ is activated by fertilization, whereas in plants the seed is activated by receiving nutrition, similar to the nutrition received by the fetus. Consequently plants have no sexuality. Generation in the plant is merely the initial growth of the seed, which is made possible by a highly purified nutrient. Of course the nourishment provided by the mother plant and by the flower in particular is much more refined than the nutrient drawn from the earth by the adult plant. The final step in the argument is to establish the location of the nutritive apparatus that feeds the seed. There are no major problems in doing this, because the apparatus can be modeled either on that which feeds the fetus in the womb or on that which feeds the newborn infant. One possibility, of which Malpighi availed himself, was to look upon the whole flower as a sort of womb containing the seed as a fetus. Another possibility was to follow Hartsoeker and compare the flower to a gland: "When this fruit begins to emerge from the opening bud, it is like a suckling infant, taking its nourishment from the flowers, which serve it as nipples, because at this stage it is still too tender and delicate to digest the nourishment that would come to it directly from the stems or leaves, which are like the viscera of the plant."[50]

Study of the flower with respect to both its anatomy and its physiology can now be contemplated. Since the purpose of the flower is to provide a nutrient delicate enough for the seed, it must contain a purification apparatus. A use was thereby found for the filaments that surround or surmount the part of the flower that contains the seed. If we follow Malpighi and Grew and compare the seed to the fetus, we are forced to take the view that the substances exuded by the "attire" (the term Grew used to denote the stamens and the pistil) are analogous to the menstrua in women.[51] Is the function of a woman's period not to purify the organism prior to gestation? The sap accordingly separates out superfluous matter in such a way that the seed receives only the nutritive principles. If we follow Hartsoeker and Tournefort, however, and compare the seed to the newborn infant, we must then assume that the filaments are "excretory ducts." For Tournefort, there was no doubt that "the anthers receive all that is not quite suitable in the food, and their valves open slightly under the pressure of the excrement."[52] To be sure, the observable differences between flowers of various species did not escape Tournefort's notice. The differences did not complicate the question of the generation of seeds, however. Flowers were assigned the role of viscera, and it was perfectly possible to alter the description of the mechanism by which they were supposed to perform their functions. For example, in plants that carry staminate and pistillate flowers on the same stalk, the former function as "kidneys." In plants where the two kinds of flowers are on separate stalks, it suffices to argue that it is the filament above the seed, the pistil, that serves as an emunctory. "I call the part of the flower that usually occupies the center or middle of the stamens the pistil.... It serves as an excretory duct."[53] Tournefort has no difficulty explaining why flowers eventually fall. When the fruit has increased in size, it no longer requires refined nourishment and can subsequently be fed by the sap that comes from the earth through the ducts of the branches. Growth of the fruit causes the ducts of the flower to be compressed until finally it falls.

Not until the second half of the eighteenth century was the agamist system confirmed with the addition of a demonstrative component. The reason for the delay is simple; before the necessary experiments could be done, a unique definition of the pollen was required. However coherent Tournefort's system may have been, it closed off the way to experimentation in advance by assigning the stamens a negligible role in some cases and an essential role in others. Their role was negligible if not nonexistent in species in which one plant bears the flowers and another bears the fruits, because in these excretion is accomplished by the pistil. By contrast, their role was essential in species in which flowers and fruits are found on the same stalk and in species in which the flowers surround the fruit, because the excretion of pollen through the stamens in these species contributes, according to Tournefort, to the generation of the seed. Obviously this meant that the stamens play two contradictory roles, as a result of which no experimentation was possible. In order to eliminate the difficulty, though, would it not suffice to substitute the question of the localization of the germ for the question of the mechanism of generation in the different species of plants? Indeed, if we follow Spallanzani or Bonnet in asserting that the seed contains the plantlet, it becomes possible to suggest a definition, albeit a negative one, of the pollen: The powder does not contain the plantlet. Although this makes experimentation possible, it is still not clear why it should be necessary. Furthermore, it will be objected that supporters of the ovist thesis, far from denying that the germ is activated by fertilization, recognized the usefulness of fertilization, at least in the generation of animals. Accordingly the application of ovism to plants would only move the difficulty back one notch.

In fact there is a very close relation between the ovist theory and an experimental demonstration of agamy. In order to shed some light on this relation, we must first call attention to a methodological problem that seems to have received little attention from historians of botany. In the middle of the eighteenth century, physiologists noted a dissimilarity between generation in animals and generation in plants. Thanks to the work of Haller on chicken eggs and of Spallanzani on the eggs of amphibians, the arguments used to support the ovist thesis were drawn from the animal kingdom and not the vegetable. Preexistence was demonstrated by Haller in chickens and by Spallanzani in frogs, salamanders, and toads. Consequently, the problem at this stage, in contrast to what had been the case in studies of generation at the end of the seventeenth century, lay in observation of the seeds: "The testimony of the senses is not favorable to the preexistence of the germ in unfertilized seeds, though it is favorable to preexistence of the germ in animals."[54] The reason for this statement was that observation of the tiny silique revealed only a spongy, homogeneous substance, a sort of jelly in which no lobe or plantlet could be distinguished. If the hope is to eliminate this disparity and prove that nature's laws are uniform, there is only one way out, to substitute experimentation for observation. Spallanzani, however, could do no more than the sexualists who kept the pollen away from the ovaries. It is clear that, ultimately, there were two possible outcomes to experiments of this type. It might be shown either that generation requires the action of the pollen, which would prove that animalculism and sexualism were correct, or that the process of generation takes place without the action of the pollen, which would settle the question in favor of ovism and, correspondingly, agamism. Thus it is clear that, in plants at any rate, proof of ovism implied proof of agamism.

In actuality Spallanzani favored ovism and hence agamism from the beginning. It is clear a priori that the demonstrative component of the theory could not fail to confirm his prejudice. Blunders were therefore inevitable. Knowing in advance what they wanted to see in the experiments, the ovists also knew what they had to do to ensure that the interpretation of the experiment would conform to their expectations. Thus, when von Sachs speaks of Spallanzani's "carelessness," he has surely got things the wrong way around, particularly in view of the fact that Spallanzani took endless precautions, for example placing the female stalks in glass vessels. We know too that Seraphino Volta reproached Spallanzani for not having himself performed the experiments that he relates. Whether or not Spallanzani actually carried out the experiments is of secondary importance, however, given his ovist presuppositions. Moreover, if Spallanzani found that dust was necessary for the fertilization of basil and garden mercury, his experiments with squash, watermelon, hemp, and spinach yielded diametrically opposite results. Well before Spallanzani, Alston had done similar experiments on spinach, hemp, and garden mercury, obtaining the same results. Although kept away from males, these species "produced fertile plants and were filled with seeds that grew to maturity."[55] Möller also found that removing the male flowers from a field of spinach and hemp did not prevent the female flowers from producing fertile seeds. Finally, Reynier found that *Malvaceae* as well as hollyhocks continued to develop after their sexual organs had been amputated: "My experiments concerning fertility without fertilization were not disappointed."[56] These experiments yielded two results. The first had to do with preformation and particularly with the localization of the germ: "Since the embryos are not part of the dust of the stamens, it must be that they are part of the ovaries, which are their natural seats."[57] The second result obviously involved fertilization: "In establishing that there are plants that are fertilized without male dust, I am attacking the modern botanists and physicists who have set Cesalpinus and all those who followed him against me, including all the most celebrated naturalists, Grew, Ray, Camerarius, Morland, Vaillant, Jussieu, Duhamel, Adanson, and Bonnet, who accepted the idea of two sexes in plants, which cooperate in the process of fertilization."[58]

No doubt what interested Spallanzani and Alston was not so much the confirmation of agamy as the confirmation of the ovist theory. Although the question of the generation of the seed was secondary, it could not be neglected. The system proposed earlier by Tournefort again became current, because the experiments that showed the dust was of no use in the generation process confirmed it. Had not Tournefort contended that there was no relation between the generation of the seed and the excretion of the pollen in species that bear their staminate and pistillate flowers on different stalks? But Spallanzani's experiments with basil, the flowers of which have both pistils and stamens, showed that the excretion of pollen contributes to the development of the seed. "In hermaphroditic flowers, privation of the dusts did not prevent the embryo from appearing in the seeds, although these seeds did not germinate in earth."[59] This result, moreover, was not incompatible with the explanation proposed by Tournefort. Had he not claimed that in flowers with both pistils and stamens the excretory function contributes to the generation of the seed by relieving the embryo of superfluous matter? Hence there is no cause for surprise if, after removal of the stamens, the seeds abort. "These parts merely act like any number of other parts of the plant, which, when they happen to be missing, impair fructification, though no one would imagine that they promote generation."[60]

The demonstrative component of the theory also served to emphasize the ambiguity in the nature of the pollen. One could well be certain that the pollen contains no plantlet. Yet the role of the stamens in the process of generation was hardly clear. Some writers claimed that the stamen plays a useful role, performing an excretory function that is necessary to the emergence of the seed. In that case, however, species in which staminate and pistillate flowers are disjoint present a problem. It is hard to see how the staminate flowers could relieve the embryo of excremental substances if the seed is located on another stalk. Other writers held that the stamen plays no useful role, there being no connection between the excretion of the pollen and the generation of the seed. In that case, however, the intimate connection between the stamens and the pistil in flowers that have both became unintelligible. Owing to this ambivalence in the excretory function, the problem of generation of the seed remained in suspense, as it were, between two irreconcilable solutions. The scope of agamism was thereby significantly reduced. At best it would have been possible, out of a concern for consistency, to apply one solution or the other to all vegetable species. No matter how little attention was paid to the pollen, it was clear that it could be looked at in two different ways. Was it not an excrement? As such, it was both useful and useless, that is, nutriment and waste, life and death. This led to a split between two opposed and, what is more, incompatible interpretations.

First of all, it could be argued that the powder plays no role in the process of generation. From this it followed that the excretory function of the stamens and the generation of the seeds are two completely distinct mechanisms. There were two ways of looking at this disparity, depending on what meaning was ascribed to the word *pollen*. On the one hand, since the powder is an excrement, it was compared with waste matter. As such, in Alston's view, it was especially noxious. "Nature has arranged for this dust to be thrown away as far as possible, for it is useless, if not injurious, to the style."[61] Consequently, excretion must be arranged so as to prevent any contact between the powder and the part that contains the seed. In this view of the matter, the separation of staminate and pistillate flowers can be interpreted as a wise precaution on nature's part. By contrast, flowers in which the pistil is flanked by stamens present a problem. In such flowers has not nature placed the seed close to the excrement, thereby menacing the former? Close examination of these flowers, however, reveals that, on the contrary, they are designed in such a way as to prevent the pollen from falling onto the stigma. In some flowers the naked eye showed that there was no communication between the flower and the seed. The position of the anther in the iris and bellwort, for example, made it "impossible for the pollen to fall onto the stigma."[62] In certain flowers the style and stamens are of unequal size; in the syngenesia and in umbellifers, for instance, the style is longer than the stamens. Finally it was observed that the two parts mature at different rates, either the style emerging before the stamens or the reverse. "The stamens of spinach, garden mercury, hemp, corn, juniper, pansies, etc., commonly throw their dust before the stigma is visible."[63]

Looking at the matter in another way, however, it was also possible to compare the pollen dust to dead matter. From this standpoint, the excretory function of the stamens no longer even enters into the picture and can be neglected entirely. The development of the seed still has to be explained, though. Following Alston, one can choose a term of reference in the plant kingdom. The bud or bulb can be chosen as a model instead of the fetus, because they contain the seeds of rudimentary new individuals. "These bulbs contain the primordial plantlet; they contain the most essential part of the seeds, because *the essence of seed lies* in the corcule, which is nothing other than *the new primordial plant.*" Comparing the seed with the bulb, one finds that no sexuality "is necessary for the production of the primordial plant, the essence of seeds as well as of bulbs."[64] The argument that the seed develops in the same way as the bulb leads ultimately, however, to the establishment of a dividing line between sexual generation in animals and asexual generation in plants. Contrasting the modes of reproduction in this way conflicts with the idea that the reproductive function gradually becomes simpler as one descends the chain of being. Accordingly in order to avoid establishing a sharp discontinuity between the two kingdoms, what needs to be done is not so much to raise the status of the plant kingdom or to lower that of the animal kingdom as to consider the intervening link of the chain. Now, asexual reproduction, which is the model on which the generation of the seed has been based, was also the model on which reproduction of many lower animals was conceived. This took care of the transition. Another advantage to proceeding in this way is that it became possible to confirm the validity of the comparison between the seed and the bulb and hence to confirm the validity of the agamist theory in plants: "If many species of animals are destitute of all the endearments of love, what should induce us to fancy that the oak or mushroom enjoy these distinguished privileges?"[65]

Alternatively it is possible to contend that the excretion of the pollen does play a role in the process or reproduction. The function of the stamen is then highly useful, contributing actively to the development of the seed. With the expulsion of the pollen everything begins. There are two ways of looking at the pollen's role, however, depending on what meaning is attached to the word *pollen*. First of all, the excrement can be held to be a vital principle. Put forward by Schelver and Henschel, this theory revolves around the idea that the powder, "in falling on the stigma, causes a sort of illness or mortification to develop there, which stops the vegetation of that part; this causes the sap to flow toward the ovules and forces them to develop."[66] Another possibility, of which Fougeroux de Bondary availed himself, was to identify the excrement with a fertilizer, that is, with a nutrient. This is probably the origin of the seemingly nonsensical idea of looking to see "if this dust might fertilize a female plant by being introduced through the roots, which would draw it in and carry it to the embryo, thereby causing the germ to develop."[67] Bonnet, who urged Spallanzani to carry out this experiment, observed that the roots are, in any case, quite far away from the flowers, and that the fecundating spirit would have to travel a long way to reach the ovary. "The shorter path from the leaves and, above all, from the petals should therefore be tried."[68] Smellie, who adopted this view, was in a position to explain why hybrids occur. Among the suspected causes of degeneration was the nutriment. If this supposition turned out to be correct, then it might well be that the dust emitted by the flowers of one species is absorbed by a plant of another species, and there was no reason why "this foreign nourishment [may not] occasionally introduce some changes in the colour, texture, or flavor, of the leaves, flowers, or fruit?"[69] Smellie and Spallanzani do not, however, say whether or not they actually tried these experiments, and if so what results were obtained.

This way of solving the problem had a major disadvantage, in that it established a sharp discontinuity between the two kingdoms in regard to the reproductive function. If pollen is a kind of food, and if, as Bonnet thought, propagation is merely evolution dependent on nothing other than activation of the germ, would it not ultimately be simpler to admit the existence of a kind of sexuality in plants? Of course Spallanzani's experiments had established that the dust was usually not necessary in generation. Bonnet did not question these experiments, but there was no contradiction between these results and the idea that "this question may be the same as the one raised by plant lice, in which I have shown quite rigorously that there is a real distinction of sexes, and yet propagation is accomplished without copulation."[70] Taking the reproduction of plant lice as a model eliminated what was disconcertingly anomalous about the few experiments that showed that pollen is necessary for generation. As for the more numerous species that propagate without the aid of male parts, obviously there were still good reasons to deprive them of their sexual organs. Even if these organs were of service only occasionally, "experiments with plant lice had demonstrated that a single action of the dusts can, though with no great likelihood, cause several generations of the same plant."[71] For Bonnet, the behavior of all plants was to be compared with an anomalous phenomenon in the animal kingdom. In plants, he believed, parthenogenesis is the rule and sexual reproduction the exception. Furthermore, there was no discontinuity between the animal kingdom and the plant kingdom, the plant louse providing the intermediary link between the two.

Before long it was on another segment of the chain of being that the reproductive anomaly came to be situated, however. No longer did it describe the entire plant kingdom, as Bonnet had thought, but rather only the cryptogams. In coordination with sexualism, agamism was modified and reworked. The objective was no longer the same. For agamists the problem was not to generalize their own system but to oppose the generalization of the sexualists' system. As a result, they fell back on defensive positions, acknowledging the existence of sexuality in plants, if not in all of them, then at least in the phanerogams. They then proceeded to fortify their new line of defense, by annexing the tiny domain of lower plants. What prevented the structure of agamous plants from being superimposed on the phanerogams was of course the fact that some "difference" stood in the way. In order for this difference to be observed, however, cryptogamy first had to be demolished. Cryptogams, it was said, were unreal, figments of the imagination not so much of sexualists as of "cryptomaniacs."[72] Indeed, for de Necker, "it is not enough to freely imagine or to draw on paper attributes that somebody or other is pleased to call flowers, stamens, pistils, fertilizing pollen, and seed. . . . The uses of all these parts must be proved by means of conclusive experiments in a large number of plants."[73]

In, say, mosses, however, the smallness of the parts prevents castration of the supposed sexual organs. But there is nothing to stop an experimenter from throwing the dust of one species onto the stigma of another in order to try to obtain hybrids. For hybrids are "the true essential character for demonstrating not only the marriage of plants but also the real existence of their semen."[74] Only two things can happen; either mosses will yield hybrids, demonstrating the existence of sexuality in them, or they will not, indicating that cryptogamy does not exist. De Necker was of course following the procedure of the sexualists, but at a later date. This point is crucial, because it made a difference which system he chose to try to verify. His choice of Linnaeus' system led him to deposit the seeds of one species on the male part of another species. Obviously he did not obtain a hybrid. His conclusion was that "if the production of new species either spontaneously or artificially tends to support sexualism, other species exist that tend no less to refute it. There are no grounds for astonishment that sexless mosses are incapable of forming new species or hybrids."[75] Because mosses are sexless, they reproduce by gemma rather than by seed. The accent here is being placed on the structural differences between the reproductive bodies of agamous plants on the one hand and phanerogams on the other. In the latter there are seeds, whose cotyledons enclose the embryo; in the former there are neither cotyledons nor embryos but rather gemmae, or "gongyles."[76] This group included not only mosses but also ferns and algae. So that "to look for flowering or sexuality of any kind in algae is tantamount to looking for a head in acephalous mollusks, for eyes and ears in jellyfish."[77]

PLANT AND ANIMAL: SIMILARITY AND DIFFERENCE

When the physiologists had completed their study of generation, one question remained in suspense. Because this function is in some measure common to both plants and animals, it was imperative to establish in what respects reproduction is similar in the two kingdoms and in what respects it is different. But there were two ways of going about this task, because the characterization of similarity and difference depended on the way in which the reproductive function was described in plants. Agamists had no difficulty accomplishing their goal. Since reproduction from the seed was based on the model of the shoot or gemma, plants were distinguished from animals by the absence of a sexual function. "The generation of vegetables is accomplished without sexuality."[78] Sexualists, on the other hand, ran into trouble, because they had chosen to base their explanation of generation in plants on the model of the sexual function in animals, which in turn is based on the act of copulation and therefore on the existence of two individuals driven by instinct. Thus sexualists faced a dilemma. If the plant and animal kingdoms are alike in respect of generation, then they are alike in general, so that it is impossible to indicate the characteristic differences distinguishing the two kingdoms. But if it is possible to exhibit this characteristic difference, then generation in plants is based only partly on the model of generation in animals and the similarity between the two kingdoms is left out of the account. This either/or logic complicated the question considerably. It is evident a priori that the problem is insoluble. It can be approached in two ways, both of which leave the botanist at a loss how to proceed and are therefore doomed in advance to failure.

To bring out the difference between the two kingdoms, it suffices to observe that most animals resemble hermaphroditic flowers in the mechanics of sexuality, the union of the sexes. This identity gives rise to a difference in the mechanisms of generation but not to a divergence in the modes of reproduction. In plants there is union between the stamens and the pistil; in animals there is union between the penis and the vulva and hence between individuals, which is copulation or mating. This contrast is related to one of the distinctive characters of plant life. "A plant is a living organic body . . . which has the faculty of reproducing itself, but without copulation."[79] This is obviously true not only of plants that bear unisexual flowers, whether on the same stalk or on different stalks, but also of bisexual flowers. "This is the only difference between plants and androgynous animals, since the latter, although equipped with the organs of both sexes, cannot use them for generation without the help of another individual of the same species, whereas plants couple in themselves without any outside assistance."[80] The difference between the plant and animal kingdoms is thus given prominence in this account, but the similarity is omitted.

Alternatively the similarity can be pointed out. In that case it is necessary to do as Erasmus Darwin did and maintain that the stamens and pistils represent males and females. The union of the sexes in bisexual flowers is then seen as a form of copulation. "The anthers and stigma are real animals. . . . They are affected with the passion of love and furnished with powers of reproducing their species."[81] Another possibility, of which Linnaeus and Bonnet availed themselves, was to shift the terms of comparison and liken the stamen to man and the pistil to woman. The result was that the relations between the sexual parts of plants were viewed as analogous to the generally observed relations between individuals belonging to the same society. Linnaeus, in fact, designated species in which male and female flowers are found on the same stalk by the admirably descriptive term "monoecious," and species in which male and female flowers grow on separate stalks by the term "dioecious."[82] In the latter Bonnet explains, "the lovers live in separate apartments in separate houses," while in the former, "the lovers and nymphs, although they keep separate apartments, live under the same roof."[83] Just as plant sexuality can be made intelligible by being modeled on a society, so too can the movements of the parts of plants be explained in terms of desire in animals and even men. The stamens feel the extreme excitement aroused by the sexual instinct. Desfontaines described the quivering of the male parts, which he said was caused by "the action of the pistil itself, which incites each stamen to orgasm, similar in a sense to the familiar orgasm that occurs in the sexual parts of animals."[84] As for the pistils, there is a certain reserve in their bearing. Desfontaines also describes the scarcely perceptible expansion of the tulip or the style of the nigella, "as if the law requiring a certain modesty of females were common to all organized beings."[85] The long and the short of this account, then, is that the plant and animal kingdoms are completely identical. The difference between them is left out.

Not until the second half of the eighteenth century do the true outlines of the similarities and differences between the kingdoms take shape. This could happen only after two avenues of research had converged. One line of research established the modalities of generation in the two kingdoms, and the other looked beyond this diversity toward a common mechanism in all living things. With Koelreuter and Sprengel, the first problem was to work out a theory of pollination. In other words the problem was to generalize what had been considered an exceptional mode of pollen transfer and dispose of the pseudogenerality of self-pollination in hermaphrodites. Curiously enough, a century separates the discovery of plant sexuality from the work on pollination. The reason for this delay is that throughout the eighteenth century the main problem was to prove the existence of plant sexuality. Proofs were therefore accumulated. Some were based on the placement of the organs in bisexual flowers, others on the existence of hybrids in unisexual species. It was further believed that bisexuality was a consequence of the immobility of plants, which made autogamy the most effective means of assuring their fertilization. As for dioecious plants, nature had no choice but to rely on an some external agency to accomplish fertilization, and the unreliability of this agent was ignored.

Once botanists were convinced of the existence of sexuality in plants, however, they could not help being puzzled that fertilization should depend on such unreliable agencies. Koelreuter gave vent to his surprise after observing the pollination of the cucumber and gladiolus by insects: "I was stupefied when I first made this discovery in one of these plants, and I saw that nature had left so important a matter as reproduction to chance alone, to a happy accident."[86] This surprise did not go without some suspicion of the teleological arguments that were all too readily advanced, without further research, to explain how it was that nature makes do with the unreliable agency of insects. Was it not more likely, suspicious minds wondered, that reproduction was actually based on a reliable intermediary? To answer the question was easy: just focus on an isolated aspect of experience, the behavior of insects around flowers. Accounts of artificial fertilizations accordingly gave way in the literature to reports of insects consorting with field flowers. Favorable observational conditions were spelled out: "a clear day and warm temperatures."[87] And thus the groundwork was laid for the demolition of the idea that a sexual mating takes place in hermaphroditic flowers. Once pollination was seen to depend on an outside agency, it became possible to investigate the properties of flowers that are conducive to its success. The flower thus became the primary focus of research into the teleology of reproduction, opening the way to investigation of the teleological interdependence of natural bodies. Koelreuter and Sprengel noted the interlacing of teleologies. On the one hand there was the insect's hunt for booty, and on the other hand there was the pollination of flowers. There are two ways of studying the relations between insects and flowers, however, depending on whether priority is given to the one or to the other. One possibility, of which Koelreuter availed himself, was to give precedence to the insect's purpose, the search for food. The animal moves about in order to feed itself. Thus Koelreuter focused his attention on the plunder, which he viewed not so much as the lure offered by the flower as the food taken by the insect. He therefore studied the nectar, showing that the bee was drawn to it because it was a sugary liquid from which the bee could manufacture its honey: "The sustenance the bee requires consists of drops of nectar hidden away inside the flower."[88] Koelreuter was not, however, interested in either floral structures or in the ways in which insects went about searching for their treasure. His investigation of pollination revolved around a description of the way the pollen was loaded, transported, and deposited on the stigmata. He also made sure that the agency was actually effective by doing the following experiment. In one location Koelreuter artificially fertilized 310 flowers, while in another place he left the same number of flowers to be fertilized by insects. The results were that the number of fertile seeds was approximately the same for both lots.

Sprengel availed himself of another possibility, namely, to give priority to the flower's purpose, pollination. In this case interest was focused on the arrangements for protecting the treasure inside the flower. From this standpoint, the nectar was not so much the insect's food as the flower's means of assuring its reproduction. Thus it was appropriate to consider the treasure hunt itself rather than the comings and goings of the insects. Before being a carrier of pollen,

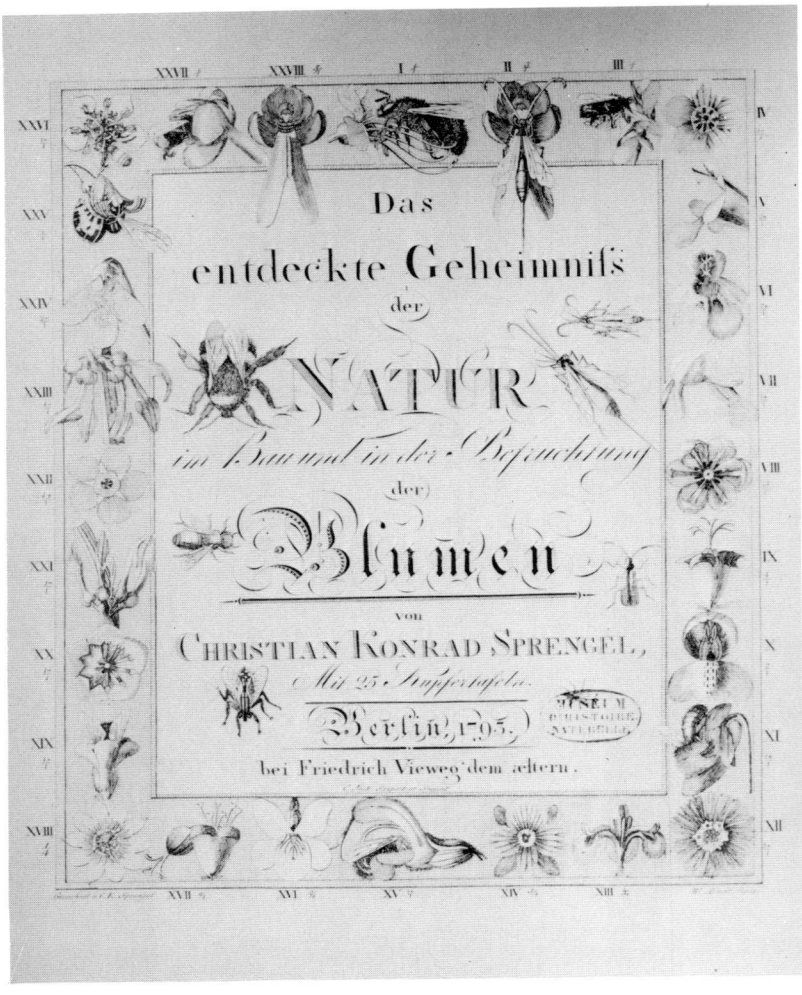

Christian Konrad Sprengel, title page of *Das entdeckte Geheimniss der Natur im Bau und in der Befruchtung der Blumen* (Berlin, 1793). Photo courtesy of the Bibliothèque National, Paris.

GENERATION

the insect is a living thing with senses of sight and smell. Only an animal capable of detecting the bait, as it were, is likely to be caught in the trap. Sprengel laid out a teleological model of explanation in his description of the five hairs situated just above the sweet droplets in the flower of *Geranium sylvaticum*.

> Convinced that what appear to be the most insignificant details have their raison d'être in the order of things established by the wisdom of the Creator, I began to reflect upon the probable utility of this villosity. After a short while, it occurred to me that the plant possesses five glands with five drops of juice inside, intended to feed the insects; given the circumstances, it seemed reasonable to me to assume that, by the dispensations of providence, these hairs had been placed so as to protect the juice against the harmful effects of the rain. . . . The hairs, which are located above the droplets of juice, hold back the rainwater and prevent it from coming into contact and mixing with the juice, just as a drop of sweat that rolls down a man's forehead is held back by the brows and lashes, which keep it from entering his eye.[89]

Not enough emphasis was placed, however, on the link between this observation and the various interpretations of the peculiar features of different floral structures. The arrangements for protecting the treasure are effective only because there exists a system of signals to show visitors the way to get to what they are after. This is where Sprengel got the idea of asking about the significance of the yellow circle that surrounds the opening of the corolla in *Myosotis palustris*, standing in sharp contrast to the sky blue color of the leaves: "What could nature's purpose in coloring this circle have been, if not to show insects the way to the reservoir of nectar? With this hypothesis in mind, I examined other flowers and found that most of them confirmed it."[90] This result in hand, a whole series of carefully made observations fell into place. Points, lines, and patches of color indicate where the treasure is located when the insect is close to the flower, while the color of the corolla does the same job but when the insect is farther away. Flowers that open at night have a brilliant corolla. In plants that have neither corolla nor nectar, nature compensates for the flaw by emitting a perfume that attracts insects.

Within the terms of pollination theory, hermaphrodites behave like dioecious plants; they can be fertilized by pollen brought to them by insects. Once this was recognized it became possible to fit what had been problematic observations into the system. Among these was the fact that in some flowers stamens and pistil mature at different times. Either the male organ appears before the female organ, as Sprengel observed in the *Epilobium augustifolium* and the *Nigella arvensis*, so that these flowers are not fertilized by their own pollen but rather by pollen gathered by insects from flowers that have bloomed more recently.[91] Or the female organ appears before the male organ, as in the *Euphorbia cyparissias*. "When the sexual parts (the anther and stigma) develop at different times, I call the arrangement 'dichogamy' for short."[92] However, he goes on, "since there are two sorts of dichogamy, they must be distinguished by different names. The first to be discovered I call male-female; and the second, female-male."[93] The other set of problematic observations concerned the arrangement of the parts of plants, as well as differences in the size of the sexual organs. Because pollination was now recognized to be accomplished by insects, there was no longer any need to invent, as Linnaeus had done, mechanisms capable of ensuring autogamy. What is more, anomalous plant structures could no longer be interpreted as arguments in favor of agamism. In short, cross-fertilization ceased to be the exception and became the rule. "Since many flowers are dioecious and since at least as many hermaphroditic flowers are dichogamous, it would seem that nature wanted not a single flower to be fertilized by its own pollen."[94]

The purpose of the insect thus derives from the task it

must accomplish. On these assumptions, however, how are hybrid plants to be explained? Animal hybrids, we know, were explained as the result of a "perversion of instinct."[95] The underlying assumption is that, normally, animal instincts conform to strict rules that prevent crossbreeding between different species. Plants, on the other hand, exhibit no such regularity of behavior. The only possible argument, which was actually made by Koelreuter, was to invoke the geographic separation between different plant species. It was part of nature's design to "transfer one plant to Africa and to assign to another its place in America."[96] Plant hybrids were then explained as the result of placing a plant in unnatural conditions. What did man do in his gardens, if not bring together species that nature had been at pains to keep apart?

Furthermore, as François Jacob has pointed out, "some of these hybrids were fertile. In their progeny certain characteristics of the parents were seen to appear through successive generations. All this was hardly compatible with preformation."[97] When the second line of research was pursued further, it was for this reason that a unified internal mechanism was proposed as an answer to the apparent diversity of modes of generation. Whether fertilization is the result of copulation (as in animals) or of pollination (as in plants), it is clear that it consists in the union of elements issuing from the male and female parts. Not only Koelreuter but also Maupertuis and Buffon made use of a chemical model. For Koelreuter, hybrids had an intermediary nature, "just as the union of an acid salt and an alkaline salt forms a third salt, which is neutral."[98] For Maupertuis, "when one mixes silver and spirit with mercury and water, the parts of these substances rearrange themselves to form a vegetation so similar to a tree that it has been impossible not to call it by that name. . . . What a miracle, if such a vegetation occurred on its own, out of our sight!"[99] Here lay the point of similarity between the plant and animal kingdoms. The distinction between pollination and copulation resolved the question of the difference between the two kingdoms in a completely satisfactory manner. The animal was distinguished from the plant by dint of its relations with other individuals, for which purpose it was equipped with senses and the ability to move about. Mating or copulation required mobile, sensitive structures in animals. Plants, however, reproduced by means of a go-between and so had no need of such an organization. Generation was accomplished through a different type of relationship in plants and animals, and ultimately it was this difference that established the distinctive features of each kingdom. This did not, however, diminish the similarity between the two kingdoms with respect to generation per se.

PLANT SEXUALITY: RESISTANCE AND DESIRE

During the eighteenth century it was noted that there was an affinity between religion and the study of botany. One of those who made the observation was Condorcet, for whom the relationship between the two was subsumed under the head of natural theology: "In reading the history of the sciences, some people have formed the belief that some scientists are more disposed to piety and others less, depending on what sort of knowledge they cultivated; and botanists, these people think, deserve to be placed in the front rank. Indeed, it seems that it is in the plant kingdom that one finds more unity of design and vision.... Observation of the plant kingdom seems to call up more forcefully the idea of a first cause, to tell us more about its boons, and to incline our soul more naturally to gratitude."[100] Botanists rated this approval because they were quite willing to accept the moral censure that fell upon unfettered gratification of the desires. Their attitude was not incompatible with pious feelings. Herborization, moreover, was the favorite activity, not to say amusement, of puritans.

At the same time, the repugnance that was felt against animals merely reflected the disgust aroused by carnal love. Anyone who took an interest in animals was suspected of bestiality. The chase, combat, and the duel were occupations suitable only for brutes. Sensitive souls were also repelled by the microcosm of insects. Did they not have to be pinned? With anatomy, moreover, the study of animals took a really repulsive turn. "What a frightful paraphernalia an anatomical amphitheatre contains, with its stinking cadavers, flesh livid and oozing, blood, disgusting intestines, terrifying skeletons, and pestilential vapors? By my word, it is not to such a place that Jean-Jacques will go looking for amusement."[101] Beyond any doubt, dissection left the observer in a dubious position: prying, breaking and entering. The animal, in short, was too close to man. As for the mineral kingdom, curiously enough it awakened the same fantasies. The reason was that, to be studied, minerals first had to be extracted by excavating into the entrails of the earth. This left the vegetable kingdom, provided that the researcher knew the proper way to approach it, and, in particular, avoided the pharmacist's approach. For it was possible to associate plants with the human body through drugs. Once again the taint connected with the flesh reared its head. "These medicinal notions are hardly apt to make the study of botany agreeable. They cause the dappled meadows and bright flowers to fade, suck the cool damp out of hedged fields, and make the green glade and the forest shade insipid and disgusting."[102] In short, anything that recalls the body and its needs is a reminder of man's subjugation.

Having said this, we can see at once why botany should have seemed a path to salvation. Whether we think of it in relation to its object or in relation to its methods, we see that it turned the mind away from images of the body and of sexuality. The opposite of the animal, the plant emancipated the observer, freed him of carnal constraints. Botany was a technique for achieving mastery over the instincts. "For men of all ages, the study of nature dulls the taste for frivolous amusements, wards off the tumult of the passions, and leads the soul to beneficial nourishment by filling it with the object most worthy of its contemplation."[103] In reality, the observation of flora met the need for severe repression: name and classify rather than dream. This intellectual activity was never degrading, because plants were understood through their forms and not their functions. Silent and peaceful, the world of plants is not of a nature to terrify delicate souls. Then too in botany, unlike zoology or anatomy, the relationship that forms between subject and object is never a source of conflict. On the contrary, subject and object coexist happily, and the herbarium or nature book extends that happy relationship into the hours of leisure. An objection has no doubt occurred to the reader: What about plant sexuality? Sex gives rise to fantasy, recalls human instincts and human identity. In other words, as has often been said, sex is the incarnation of spontaneity and anarchy; sex is considered aggressive, even subversive. It would therefore seem difficult to reconcile the way sex was conceived with puritanism. Apparently, though, sex in plants did not always awaken fears. What was it if not the pretext for a secret revenge, the key to an order, at least within the system?

Disgust with sexuality was accompanied by a secret attraction that made botany inherently ambiguous. Anyone who gathers plants must reconcile the acceptance of a moral censorship with refusal to abide by certain prohibitions. Herborization is an activity that is both innocent and guilty. It is guilty because the botanist rediscovers what he is pretending to flee. Nature must therefore reflect the image of a sublimated beauty: "Trees, shrubs, and plants are the ornament and finery of the earth, which nature has enlivened and clothed in her wedding gown."[104] Herborization is innocent because it gives its sanction to imaginary gratifications. For Bonnet the pleasure was untainted. While reassuring his female readers he awakened their curiosity: "This agreeable part of natural history, the study of which every day brings delights to ladies, may interest you in a still more lively way, when you learn that by analogy you will be able to find in plant love pure and delicious enjoyment."[105] What was the result of contemplating the loves of plants, if not to bring into being an economy of desire? No longer was desire thwarted by taboo, now that it could find gratification in a world of rustic imagery. Among botanists, moreover, the sexualists were not exempt from puritanism. There are good reasons, therefore, to question the explanation usually given in histories of botany for the opposition between sexualism and agamism. These explanations are of two kinds, psychological and philosophical.

Psychology first: we are told that agamism, and hence antisexualism, was the result of a reaction against sexuality. "Alston's attempt at refutation was motivated by outraged propriety."[106] At the same time it is implied that sexualism was favored by men of more open mind. In fact, all botanists, whenever they confronted the problem of representing the sexes, were obliged to observe the same requirements of modesty. Linnaeus remarked that the sexual organs, which are "considered as almost shameful in the Animal Kingdom, are almost always hidden by nature." It may well be objected that Linnaeus was projecting a cultural interdiction into the animal kingdom in order to heighten a contrast. For he goes on to say that "it is agreeable to recall that the genital organs of plants are exposed to the view of all in the Plant Kingdom."[107] But the discrepancy is not what it seems to be. Neither in the plant world nor in the animal world can the taboos be violated. There are, of course, two ways to combat the shocking image of the plant exhibiting its sexuality. The agamists availed themselves of one of these possibilities by mocking the idea. "It will not be expected that I should mention those parts of Linnaeus's reasoning which are derived from analogy. In many instances, he has pushed analogy so far beyond all decent limits, that it becomes truly ridiculous."[108]

The other possibility was to strike out the unsupportable image by stating the reason why nature uncovered the parts of plants. This was what the sexualists did. Surely their procedure had the effect of eliminating the sin of exhibitionism on the part of plants; by transfiguring the flower, the sexualists actually restored the privacy of the sexual organs. One way of doing this, followed by Rousseau and Linnaeus, was to identify the flower with woman: "The sweet fragrances, the lively colors, the most elegant shapes seem to vie with one another for the right to hold our attention. One need only love pleasure to abandon oneself to such sweet sensations."[109] Conversely, woman is flower—that is, sex: she dresses and ornaments herself the better to unveil her essence. Thus the difference between the nude and the clothed body was denied. The pleasure arising from the contemplation of this beauty was short-lived, however, for in the identification of the flower with woman, the opposition between nature and culture had already reasserted itself, with the end result that the difference between the two was heightened. The sexual organ is dark and evil-smelling. By contrast, the flower is fragrance and ornament: "Much as the genital parts of all animals have a strong and repulsive odor during the season of rut, so do flowers, or the genital parts of plants, exhale an odor, which, though quite varied in different plants, is most of the time very sweet. This is why man himself imagines he is drinking in nectar with his nostrils."[110] A second way of proceeding was actually only a variant of the foregoing. This was to compare flowers with women's faces, as Rousseau and Gleichen did. Flowers then exhibit "charming and gracious structures."[111] Conversely faces are flowers, that is to say, sexual. The veil that covers a woman's face serves to unveil her essence. Top becomes bottom, in other words—or, more precisely, neither top nor bottom: "The animal has only to be turned upside down to put its genital parts in the same position as those of plants."[112] Again, the difference between nature and culture has been obliterated and, again, the pleasure in doing so is only short-lived because the inversion is unreal and the denial of the difference between high and low only reasserts itself more strongly than before. The high and the low, the noble and the base cannot be superimposed on each other. The flower exhibits not its sexuality but its physiognomy. It bears the stigma of culture no less than a vulva. "There is no rarer rapture or ecstasy than that which I felt each time I observed the structure and organization of a plant and the interplay of the sexual parts in fructification."[113] Compared with this, the union of the sexes was for Rousseau a confused notion, which he says "revealed itself to me only in a hideous and disgusting form."[114]

In addition to the psychological theme, histories of botany have also harped on the philosophical differences between sexualism and agamism. The two approaches are said to have diverged because they were premised on different choices of Christian theological themes. The implication is that these themes structured the various representations of generation in the plant kingdom. One such theme was the concept of natural economy, which gave due emphasis to propagation. Those who chose this theme naturally recognized the existence of plant sexuality. What is more, this recognition could be fitted into a broader and more ambitious perspective. "If the sexual theory be not admitted, no satisfactory view can be given of these parts, and we cannot discern another useful hypothesis; in this branch of science, the naturalist's main task is the following, namely to assign the true causes of existence of natural objects and to explain the workings of nature, as well as to demonstrate the wisdom, power, and goodness of the creator."[115] On the other hand, there was the idea of the chain of being and its correlate, the hierarchy of functional principles. Taking these into account obliged the botanist to deny the existence of sexuality in plants, because animals required a faculty of sensitivity in order to fulfill the sexual function. Plants clearly lacked sensitivity. "However correct it may have been from a philosophical point of view to regard the stamens and pistils as corresponding in their respective functions to the sexes in the animal kingdom, it should not have been forgotten that in animals this process is voluntary, whereas in vegetables, notwithstanding all the ingenuity the ancients and the moderns have shown in defending the existence of a sentient principle, we are not yet justified in attributing this process to anything other than what we are accustomed to call a mechanical cause."[116] These very rational arguments, borrowed from Christian theology, lead with hindsight to the suggestion of a way of justifying sexuality in plants with a very specific intent, namely, to conceal resistances originating at quite another level. The nature of the repression becomes clear if we turn the argument around. In fact, the same standards of censorship applied to both agamists and sexualists: investigation of generation was neither to excite the imagination nor to disconcert the faculties of reason. Thus the arguments of natural economy and the idea of the chain of being were two avenues that physiologists could travel to the same end, the reduction of generation to a mere mechanism.

One way of denying sexuality was to place it in a position of prominence, as the sexualists did. Sexuality was a boon to living things and lay at the root of their harmonious coexistence. In one respect, sexuality was exalted, the better to challenge its status as an autonomous process. The sexes were said to be instruments of a purpose that transcends or, rather, outlives the individual, the perpetuation of the species. On the other hand, the positive function of sexual taboos was emphasized. The repression that stands in the way of the gratification of desire is not a source of dissatisfaction or pain but rather an aid to well-being. It is a liberating constraint, accomplishing the transition from nature to culture. Many advantages followed in consequence, for both individual and offspring. Reason in man and instinct in animals could not help but ally themselves with a desire that respects the law, the law of endogamy in beasts and monogamy in human beings. As to the pleasure inherent in the sexual act, in human beings it was perfectly legitimate, provided its source was copulation between man and wife, while in animals it was nothing less than a form of reward. Such a conception of sexuality was neither shocking nor astonishing, for ultimately sex was merely a mechanism. Thus no danger was involved in allowing plants to partake of sexuality. Drained and emptied of all subversive content, sexuality in plants was modeled on permissible human behavior, respectable marriage. The leaves of the flower "serve solely as a nuptial bed, which the great Creator has arranged in so magnificent a way, adorned with such noble curtains, and perfumed with so many sweet fragrances, that the solemnity with which the bridegroom celebrates his marriage there is all the greater. Once the bed has been made ready in this way, it is time for the groom to embrace his cherished bride and offer her his gifts."[117] And this was not all. Plant sexuality even became an image of piety. It was, in fact, an inducement to meditation, and meditation led to glorification of the creator. "When one thinks about the details of this activity, about the prodigious number of relations and combinations necessary to its success, one cannot fail to be struck. How many millions of similar activities take place in fields of limited size. How many millions upon millions of relations and combinations enter into these activities. What art, what contrivance, what steadfastness, and what intelligence these tiny objects reveal to us."[118] The more complex the activity is, the more it attests to the unity and coherence of the divine plan. God is an artist and botany a branch of aesthetics. From plants to man himself, one admired the various ways in which generation was accomplished, and through them the skill of the Creator. Furthermore, the idea of a hierarchy of living forms had its natural place in the context of the natural economy.

Alternatively it was equally possible to repress sexuality by denigrating it, as the agamists did. Sexuality was then held to be the source of all evils and, in particular, of the discord prevalent in the relations among living things. On the one hand, the constraints imposed by sexual taboos were emphasized. By preventing unfettered gratification of desire, repression scarcely enhanced well-being but rather engendered suffering by forcing living things to forgo their pleasure. This gave rise to aggressiveness, to thwarted passions, and to pain. Pleasure could be obtained under these conditions only by violating taboos, so that it became synonymous with licentiousness. On the other hand, sexuality had to be contained; the outpourings it was apt to provoke had to be dammed up. Man was therefore endowed with reason and animals with instinct. Monogamy (in human society) and endogamy (in the animal kingdom) would be impossible without these governing principles. If sexuality existed in plants, it ought to reflect this conception. Since this was not the case, the existence of sexuality had to be denied. If pleasure could be had only by breaking the law, pleasure was a vice and therefore damnable. In certain plants, however, several stamens gathered around a single pistil, as in polyandry. In others flowers were hermaphroditic, or males and females were found consorting indiscriminately, as in polygamy. This drew Siegesbeck's wrath: "Who would instruct young students in such a voluptuous system without scandal?" Furthermore, naturalists who emphasized the dependence of sexuality on reason in man and on instinct in animals were forced to admit the impossibility of establishing analogous relations in plants. Sexuality without intelligence signified perpetual violation of the laws of man, particularly the law of monogamy. "What man in the world will ever believe that God Almighty should have introduced such confusion, or rather such shameful whoredom for the propagation of the reign of plants."[119] Sexuality without instinct signified violation of the rule of endogamy: "Consider the consequences of such an arrangement. Is not this to make Nature operate against her own intentions? Nature intends that plants should multiply and perpetuate their kinds; but the sexual hypothesis makes her take the most effectual measures to prevent that intention, and to introduce universal anarchy among the vegetable tribes. Were this theory true, the whole vegetable kingdom, in a few years, would be utterly confounded."[120] Another possibility was to give preeminence to the hierarchy of functional principles, while holding fast to the arguments of natural economy. What were flowers for, if not to provide insects with food? "We see bees leaving flowers with their dust, and it is not improbable that they use it to feed themselves."[121] The economy of society, which, though conceived of as natural, was in fact rational, transmuted excrement into gold; similarly, pollen, the excrement of plants, was transmuted into honey.

Finally, it was possible to invest statements about plant sexuality with different kinds of values. They could serve, first of all, as protest. Bonnet compared what he saw in plants with the human relations he observed in his own society, and called attention to illegitimate mating in plants, where this was a natural relationship. No more than this was needed to make moral what seemed, at first sight, immoral in both plants and humans. From here it was but a short step to a denunciation of the hypocrisy of bourgeois morality: "This pleasant coquetry, against which somber moralists and absurd husbands rail so crudely and bootlessly, is authorized by nature. Wise nature, which always acts for the best, as Pope, Leibniz, and Pangloss tell us, provides abundant examples of plants that are faithful to felicitous instinct."[122] Men, too, once lived in a golden age like that which plants enjoy, but that time is past: "Nature, in that age of innocence, led a shepherdess to a band of brothers and through her made them all happy.... O happy time! ... Now, all is changed; and thanks to life annuities, each man loves himself only; only the flowers have held on to the feelings of an earlier age."[123] Besides protest, another value could be associated with plant sexuality—repugnance. For what society condemned was brought out into the open in the plant kingdom. What do we see in plants but endless violation of the laws that govern human society? "For anyone blessed with delicacy of feeling, these continual nuptials, in which monogamy, the basis of our morals, our religion, and our laws, is replaced by a licentious polyandry, are unbearable to watch."[124]

It should now be clear where agamism and sexualism came into conflict. Once the decision was made to compare the seed with the egg, the way was clear to the introduction of plant sexuality. From the outset, there was a choice to be made. Either the seed is similar to the egg, or it is different. To decide one way or the other, the comparison had to be fully worked out in detail, but with attention being paid only to the most obvious signs, sterility or fertility of the seed. The seed and the egg were found to resemble one another, and this led to further development of the implications of their similarity. The existence of a sexual function in plants could then be contemplated. The reproductive apparatus of the plant was modeled on that of the animal. Botanists looked for structural and functional similarities. The agamist theory, on the other hand, resulted from a decision to compare the egg to the fetus. Models were taken from animal physiology, leading to an identification of parts of the flower with parts of the female that play a role in the preparation of food for the fetus or newborn infant.

Sexualists experimented in order to test hypotheses. When Spallanzani or Alston did experiments, they always interpreted the results in such a way as to support the agamist system. The concept of pollination, developed by the sexualists, led to complete reformulations or reconstructions of the mechanism of fertilization in the various plant species. The agamists, on the other hand, were unable to give a unique definition of pollen and therefore amassed a series of contradictory explanations of the fertilization process.

Because agamists held that reproduction in plants was asexual, they were bound to maintain that the sexual function was a distinctive feature of animality. Because sexualists believed in the theory of pollination, they were bound to contrast sexual reproduction in animals with sexual reproduction in plants. But it was the manner of fertilization, not the mode of reproduction as such, that was different in the two classes of organized beings. Furthermore, the sexual apparatus of animals is more complicated than that of plants. The distinctive difference between plants and animals lay, according to the sexualists, in the relations between individuals and not with the existence of sexuality as such.

Finally, the same system of taboos applied to both sexualists and agamists, and both exhibit characteristic sorts of resistance in dealing with problems of generation. In one case, social taboos structured the perception of the sexual phenomenon in plants. In the other, the resistance was less elaborate but equally effective, because the notion of sexuality was simply mocked and treated with derision. It is true that discussions of generation in plants also expressed fantasies and desires. As for natural economy and the chain of being, it is clear that both of these conceptions affected sexualism as well as agamism. For the sexualists, propagation was an essential part of the natural order, but they emphasized the variety of mechanisms for accomplishing generation. Agamists stressed, first, collaboration between the two kingdoms (pollen is the food of insects) and, second, the diversity in the modes of reproduction (sexual in animals, asexual in plants).

4 Movement

Before botanists could notice the mobility of the parts of plants, they had to be familiar with the problems surrounding nutrition and generation. Thus it was not until the second half of the eighteenth century that phytodynamic phenomena began to be studied. From the outset, studies of movement revealed a discordance between plants and animals. Animals exhibited irritability and the complex mechanisms associated with muscular motion. Plants, by contrast, were very simple and homogeneous; no organs that resembled muscles, nerves, and brain could be seen. What mechanisms could possibly be cited to account for plant motion?

The first step in investigating phytodynamic phenomena was to choose a model in the animal kingdom on which the mechanisms of plant motion might be based. The choice of model was a delicate matter, because there was no way to be sure that any particular choice was the right one to explain a specific type of plant motion. There was a very simple way around this difficulty, however. If plant motions and animal motions were compared before the choice of model was made, the choice would be that much less arbitrary. At this point a further dilemma arose. Botanists might carry out this comparison by looking directly at the motions themselves, in which case the properties and functions known to exist in animals would be looked for in plants. Or they might argue that animals and plants are alike in using motion for similar purposes, namely, to preserve and perpetuate the species, but not necessarily in the modalities of motion. Thus plant motions might be governed by a mechanism quite different from any of the mechanisms known to be at work in animals. Those who chose the first alternative spoke of "plant actions," while those who chose the second spoke of "plant mechanics."[1]

As usual, histories of botany draw a sharp contrast between botanists who based their work on analogy and botanists who used observation and experimentation to investigate plant life. The former are supposed to have believed that plant motions must be exactly like animal motions, and this belief is supposed to have misled them into finding nerves and muscles in plants. Their rivals engaged in experimental research and are consequently supposed to have fared better. Charles Bonnet and Erasmus Darwin went wrong, we are told, because they relied on analogy. Not only did they confuse the two kingdoms, but, as men guided by religious ideas, they could not have done otherwise—so runs the argument—than to attribute sensitivity to plants. By contrast, experimentalists such as John Lindsay and Thomas Knight are supposed to have undertaken experimental investigations of plant motion. In consequence it is argued, they succeeded in making an unambiguous distinction between animals and plants; furthermore, by refusing to acknowledge the existence of sensitivity in plants, they presumably kept faith with Cartesianism. The crux of the question does not lie in these conflicts, however, but rather in the alternative that made it possible to resolve the same problem in two different ways, that is, to construct two different mechanisms of motion in plants.

PLANT ACTIONS

In the wake of studies of nutrition and generation, there emerged new objects of analysis and even a new series of functions: "The motions of the leaves, and parts of the fructification, which are very similar to those of animal muscles, constitute in vegetables the *animal functions*."[2] Irritability was of course a very useful property to have if one wished to attempt a mechanistic explanation of plant motions. It was not until 1755, however, that Haller used the concept of irritability to explain how muscles work: "I call 'irritable' any part of the human body whose length decreases when touched with some force by a foreign body."[3] Thus it became possible to carry the concept of irritability over into plant physiology only in the second half of the eighteenth century. This transfer was based on an argument by analogy. Like irritated muscles, stamens show signs of agitation. In the same way that the irritability of a muscle comes into play when a chemical or mechanical agent is applied to it, the anthers of the centaury become rigid when a stimulus is applied to them. Because of the similarity in the observed motions, the existence of an identical cause could be inferred: "Allow me to call this contractile power 'the irritability of flowers,' a quality that has perhaps never before been so clearly demonstrated as now, at least in relation to the fibers of muscles. This name is quite adequate, provided it is used only to denote a power that mechanist authors themselves do not claim to understand."[4] Attributing irritability to the stamens obviously meant that the similarity in the observed effects was being taken as a sign of a real affinity between the physiological mechanisms.

In the eighteenth century the fiber was still seen as the anatomical and functional component not only of muscles but also of nerves and tendons. "The origin of the term 'fiber' must be sought in an image that was used to explain the functions of these parts. Since the time of Aristotle, animal motion had been explained by comparing the articulated parts of the body to hurling devices: muscles, tendons, and nerves pulled on bony levers like the cables of catapults. The fibers of muscles, tendons, and nerves corresponded exactly to the plant fibers of which the ropes in the catapult were made."[5] The fibrous nature of the stamens led to their being identified sometimes with fibers, sometimes with ropes. Now, because of the similar nature of the fibers in stamens and animals, it was assumed that the functions were also the same. In other words, the term *fiber*, which the ancients also used to denote the lines that can be seen on the backs of leaves, returned to its source in the plant world, but now endowed with physiological significance. This explains the identification made by dal Covolo, with apparent justification: the filaments of flowers are muscles.[6] The only point that remained to be cleared up was where in the stamen the irritability is located. One way to determine the site was to proceed by a process of elimination. After removing the petals of *Berberis*, Smith grazed the anther and the outer part of the filament with a feather; he observed no movement. By contrast, if he touched the inner part of the filament, the anther moved toward the stigma. "It is clear that the movement of which we have just spoken is due to a high degree of irritability in the part of the filament that is attached to the germ, which contracts when touched. This part becomes shorter than the other and falls back on the germ."[7] Another possibility, of which dal Covolo availed himself, was to show that the power to contract is inherent in the very structure of the filament. In the same way that physiologists had observed signs of irritability in detached organs (Hoffmann had seen a heart go on beating outside the body and Haller had seen severed intestines continue their peristalsis), dal Covolo observed stamen fragments that twisted like worms at the slightest prick. With this method it was possible to establish that the tiniest portion of the stamen is irritable and that the property of irritability does not reside in one particular part of the organ but is diffused throughout.

Because the motions of the sexual organs were held to be of the same nature as animal motions caused by irritability, observations likely to confirm this theory were accumulated. Haller had observed that animals of all ages exhibit irritability and that it persists after death. Accordingly Smith maintained that "this irritability is noticeable in stamens of every age, and not only at the time when the fertilizing dust or pollen is ready to be released.... In a number of old flowers, regardless of whether the petals are still attached to the stamens, or ready to fall, or already fallen, wherever the stamens remain they always retain the same irritability."[8] Haller, however, had also noticed that irritability seems to reside in the gelatinous parts of the fiber and that it is destroyed if they dry out. Gmelin therefore argued that "plant irritability, like animal irritability, manifests itself only in the soft parts. It decreases little by little, as these parts lose their flexibility. It disappears for good when they have completely dried out."[9] Finally, the stamens of various flowers were examined for signs of irritability just as similar organs in many animal species had been examined. Smith found it in *Cactus tuna*; Gmelin in *Orchis*, knapweed, and thistle. And Bonnet stated that irritability exists in flowers of all species: "Once it has been found in the flowers of so large a number of plants, it becomes rather probable that it also exists even in plants where we have not yet been able to discover it."[10]

Haller had also shown, however, that the irritable parts are many and diversified. Apart from the genital organs, whose irritability has something unique about it, the property had been found in muscle fibers, in the thoracic duct, and in the lacteals. But in plants irritability had been observed only in the sexual parts. Should irritability therefore be attributed to the stamens alone, which would mean giving up the idea of establishing complete conformity between plants and animals? Naturalists preferred to assume that other parts also had the property, which would ensure similarity between the two classes of organized beings. This attitude was all the more readily accepted because irritability was quickly taken to be the basis of a new system of animal economy. A choice was necessary. Either irritability had to be limited to the vascular system in animals and plants, or it had to be made into a principle governing the whole range of life functions. One possibility, of which Bonnet and van Marum availed themselves, was to maintain that the motion of fluids in animals required the vessels to be irritable. Moreover, since the flow of the sap was analogous to the flow of the blood, the existence of a similar property in the ducts of plants could be assumed by analogy: "It is highly probable that the motion of the fluids in plants should be attributed to a secret action of their vessels. . . . It appears that the vessels of plants must alternately expand and contract, and that in this way the fluids contained within the vessels are propelled from one place to another."[11] For Bonnet, too, "the secretions of plants, like the secretions of animals, require that there be some secret action in the vessels. . . . Might it not be that this action depends on the same force that animates the sexual parts? Might there not be some analogy with the action of the vessels in animals?"[12]

Irritability of the vessels is of course hardly obvious in plants. Where might such irritability reside? In which ducts was it to be located? One possibility was to take the structure of the higher animals as a model and to assume, in the absence of proof, that the vascular system of plants is similar to that of perfect animals. It was equally possible, however, to invoke the structure of the lower animals and establish a mechanism similar to the one that exists in insects. The first solution to the problem was the one chosen by van Marum: "Irritability, that is, the faculty of muscle fibers to shorten when irritated, is known to be the cause of the contraction of arteries and veins, which to that end are enclosed in muscular membranes composed of transversal fibers. Are the vessels of plants, too, really provided with such muscular or irritable fibers? Their smallness does not permit us to see that this is so, even with the best microscopes."[13] The second solution was adopted by Bonnet, who referred to the *Traité anatomique de la chenille* (*Treatise on the Anatomy of the Caterpillar*). Lyonet had brought to light the muscles and membranes that surround the spiral strip of the caterpillar's tracheae. "The tracheae of plants might also be provided with membranes, and these membranes might be some sort of muscle possessing an irritability in keeping with the nature of plants. . . . These muscles might influence the movements of the stems and leaves, as well as of the sexual parts."[14] No matter which model was chosen, however, neither van Marum nor Bonnet was able to establish any sort of localization of the property, because of the extremely small size of the vessels and tracheae. The whole theory remained conjectural.

Based on a model borrowed from the animal kingdom, the idea of plant irritability was apparently put forward as a hypothetical construction. One way of testing "whether the hypothesis that irritability is the cause of the contraction of the vessels in plants can be supported by any experiment seemed to me to be the following: namely, to investigate whether the same cause that destroys the irritability of animal muscle fibers, with the effect of halting the contraction of the blood vessels, also halts the contraction of the vessels in plants."[15] The existence of plant irritability was in fact not being questioned, so that the interpretation of the experiment merely reflected a presupposition. Indeed, it is clear a priori that the proposed demonstration could not but confirm the existence of irritability in plants. There were two ways of destroying the irritability of vessels and halting the flow of sap. One procedure was to use electricity to suppress irritability; this was tried by van Marum. The irritability of muscle fibers can be destroyed by passing an electrical discharge through them; similarly, the irritability of the vessels in the branches of the euphorbia can be destroyed by causing an electric current to flow through them. "I observed that all the branches and stems of these plants that had conducted the electric ray or current for a period of twenty to thirty seconds yielded no further sap whatsoever from their wounds when they were cut."[16] The vessels had thus lost the faculty to contract and expel the sap they contained. By contrast, in a neighboring branch, which had been sectioned but received no electrical discharge, the sap flowed. The flow of the fluid had to be the result of the spontaneous contraction of the vessels. Another possibility was to destroy the irritability of the plant ducts as Coulon did, by applying a chemical solution. Applying a solution of alum or sulfuric acid to the wound of an animal causes the flow of blood and other fluids first to diminish and then to cease altogether; similarly, applying these solutions to two sectioned branches was enough to halt the flow of sap. A third branch, used as a control, allowed sap to flow for several hours. "This difference cannot be attributed to the mechanical action of these substances, because the solutions were much too dilute to produce this effect on a body devoid of life or even on another part of the living plant; thus it may be assumed to possess a considerable degree of irritability."[17]

From a different angle, there was also the possibility, of which Brown and Girtanner availed themselves, of arguing that irritability is a vital principle underlying the various vital functions. It was then called "incitability," a faculty of organized bodies that rendered them capable of being affected by outside agents. Since life depends on incitability and plants show signs of life, it may be assumed that the same principle is active in plants. Furthermore, as

Brown observed, "this faculty extends to everything that has life and consequently belongs to plants."[18] Unlike Brown, Girtanner attempted to localize the property of irritability and even distinguished three kinds of irritable fibers, "The straight fiber, which is found in the muscles of animals, in leaves, in stamens, and in several other parts of plants; the spiral fiber, which one finds in arteries, veins, lymph ducts, intestines, and, in general, in all cylindrical and conical ducts and muscles in animals and plants; and the circular fiber."[19] Girtanner was, further, in a position to explain why the sexual parts of plants drop off: their exhaustion is total. "The irritability of several insects and most plants is irreparably exhausted by the stimulus of propagation of the species, so that they die the moment after the act of generation is complete."[20] The powers that stimulate life are heat, air, moisture, and light. There are two unnatural states of incitability, however, produced by the relation between the stimuli and the faculty of incitability itself. Disease, in fact, is essentially due to a shortage or excess of stimulants. By contrast, health depends on the action of the various stimulants being correctly proportioned to one another. If good health is defined as the fulfillment of all functions in an agreeable, easy, and regular manner, Brown contended that in any event "*agreeable* was a term that could be applied only metaphorically to plants."[21] Not all naturalists made such a distinction.

Indeed, many physiologists did not hesitate to endow plants with sensitivity. Erasmus Darwin, for one, distinguished, as Haller had done for animals, the sensitivity of the fibers from their irritability in that actions due to sensitivity were supposed to be preceded or followed by feelings of pain or pleasure. This distinction is obviously based on the assumption that the similarity between plant and animals motions is a sign of a real affinity between the physiological mechanisms responsible for them. That is not all. Erasmus Darwin claimed to be able to tell when such feelings existed by noting the similarity between the motions that plants make in certain circumstances and those that we ourselves would make if placed in the same situations. When a leaf of *Mimosa pudica* is injured, it quickly closes up. "Does this not prove, that there is a brain or common sensorium, where the nerves communicate in some part of this bud or leaf? . . . The disagreeable sensation is propagated from a part to the whole, and causes the actions of some distant muscles, in the same manner as I draw away my hand when my finger is hurt."[22] We feel hurt and pain when we are hungry, thirsty, or cold—in other words, when we lack food, water, or heat. The absence of nutritive substances affects our senses. The want of stimulus produces the pain. Exactly the same thing occurs in plants. "This leads us to a curious inquiry, whether vegetables possess any organs of sense? Certain it is, that they possess a sense of heat and cold, another of moisture and dryness, and another of light and darkness; for they close their petals occasionally from the presence of cold, moisture, or darkness."[23]

If, however, one ascribes characteristically human feelings to plants, as Darwin did, then it is natural to elevate the plant kingdom to equal status with the animal kingdom. Equality between the inferior and the superior is incompatible, however, with the idea of a continuous chain of being. There would seem to be no way around this difficulty except by overcoming anthropomorphic prejudice. At the same time, one had to avoid going to the opposite extreme of denying that plants exhibit any sensitivity at all on the grounds they show no signs of having feelings like those seen in animals. This position leads to the same sort of difficulty as the preceding one. It is just as incompatible with the chain of being to insist on too sharp a distinction between plants and animals as it is to insist on too close an identity. "We have observed that, in nature, everything is graduated or moderated by degrees. We therefore cannot locate the precise point at which feeling begins. It may be that it extends as far down as plants, at any rate as far down as those plants that stand nearest the animals."[24]

Between these two extreme and equally problematic attitudes, however, a third course was open to physiologists, a course that led some of them to imagine that plants possess instincts. This required no more than a shift in empirical focus. Rather than look first at the motor reactions in plants, which are so different from the motor reactions in animals as to make comparison dubious, it might be better to focus exclusively on the purposes of specific actions in both plants and animals. Looking at plant motions in this light is of interest because it allows an analogy to be drawn with animal life that is now safe from any possible challenge. In animals, actions and reactions are not only variable but also specific to the organs that exhibit them. In spite of this diversity, these actions are always signs or expressions of feeling. Pleasure impels the animal to seek out what it needs to maintain itself; pain impels it to flee whatever might hamper its achievement of that end. If all motion is ascribed to such purposes and it is argued that plants sustain themselves, grow, and multiply, it is possible to argue that "the power of motion of which they are capable is nonetheless appropriate to their special nature and contributes to their well-being. Thus it may be inferred that plants are endowed with instinct and therefore with sensation."[25]

The argument that plant actions depend on some kind of instinct led physiologists to interpret all plant motions as adaptations to the environment. Some motions were connected with nutrition. In the early stages of vegetation, the rootlet and plantlet can be seen to move in seeds that are sown upside down. Each part bends so as to grow in the direction appropriate to its function. In adult plants roots can be seen to grow around obstacles and toward moist, rich earth in preference to dry, sandy soil. Roots will even seek out a sponge filled with water. Then there are other motions associated with generation. The petals close, for example, in order to protect the extremely delicate sexual organs and prevent the emission of pollen due to rain showers. Finally, there are the motions of the tendrils of climbing plants. "These examples of an instructive economy in plants have been chosen from among subjects that we observe every day. However, plants from hot climates, which we know tolerably well, would probably provide better illustrations of this power of animality as yet unknown."[26]

If attention is focused on the purposes of plant actions, the specificity of plant needs will also be emphasized. The actions of plants are obviously wholly adapted to the peculiar nature of each plant and carefully proportioned to its needs. Accordingly it was not thought surprising that the most careful anatomical research had disclosed no part of the plant similar to the organ that is the seat of feeling in the higher animals. Because the actions of plants differ from those of animals, it is highly probable that their structures also differ. Plants probably have no brain or nervous system, but "it would be quite presumptuous to assert that no other substance can perform these nervous functions in other living beings except that which we regard as appropriate to the purpose in man."[27] Arguments in favor of the existence of a nervous system in plants were based on observations of the lower classes of the animal kingdom, especially worms. "Plants are supposed to have no sensation, because in the vegetable system no nerves are detected; but is not sensation perceived in all the intestinal worms, in which also nothing like nerves can be found?"[28] To this it was usually added that it never occurred to anyone to doubt the existence of respiration in insects and fishes merely because they have tracheae and gills rather than lungs. Future research might well exhibit what could not yet be seen for want of powerful enough instruments.

PLANT MECHANICS

There was, during the classical era of botany, another way of looking at the question of plant motions. In this case, the analogy was based not on animal instincts or sensitivity but rather on the end to which these faculties were a means, namely, the preservation and perpetuation of the species. Since animal motions were viewed in relation to this twofold end, it was possible to infer, reasoning by analogy, that the purpose of plant motions was the same. "These are forces that result naturally from their respective organization, and which nature has granted them for their preservation and reproduction."[29] The point of resemblance between plant and animal motions was thus said to lie in the purposes to which those motions were put. It was possible to study the mechanisms that govern the motions of plants without prejudging their nature. There are only two possibilities: either plant parts resemble animal parts in that their mobility depends on some kind of instinct or sensitivity, or plant motions, though purposeful, are explicable in terms of the laws of mechanics. To decide one way or the other, animal and plant motions must be compared. Furthermore, since the point of the comparison is to detect similarities or differences, the signs indicating the presence of feeling and instinct must be precisely established at the outset.

In studying animal motion, not only must the role of the environment as a source of stimuli be considered, but the orientations of motions that depend on instinct must also be looked at, because "we see that the law has been written on each object, and three incentives have been held out to provide for Generation, Nutrition, and Preservation: *Pleasure*, *Hunger*, and *Pain*."[30] In plants these incentives seemed to be lacking. For one thing, pleasure would not be of any use to the plant, because the fertilizing material is carried by insects and wind. For another, plants know nothing of hunger, for their roots strike down into the earth. They therefore have no need to look for food. Finally plants apparently have no sense of impending danger. What good would it do them to be warned of danger, if they cannot flee it or ward it off? Plants do of course exhibit a series of motions on which studies of nutrition and generation had shed light. For one thing, research into the uses of leaves had led Bonnet to argue that the upper surface of the leaf serves as protection and the lower surface for absorption. An analogy was thus drawn between the leaves and the roots. The roots always grow downward into the earth, from which they draw the plant's nutriment. This reasoning probably suggested the following question to Bonnet and with it a new object of research: If one turned a leaf upside down, could it right itself and regain the position most appropriate to its absorptive function? Studies of generation had revealed motions not only of the sexual organs but also of the petals. The corolla protects the pollen from moisture just as the leaves protect the fruits. The motions of the petals therefore help to preserve and perpetuate the species. When teleological criteria were introduced, greater prominence was given to relations between the vegetable economy and the environment. What might cause the leaves to seek the dew and the corolla to close at night, if not the moisture in the environment?

Two further arguments of a biological order were advanced in addition to the foregoing. These pertained to the relations between man and other living creatures. Because of the first of these two arguments, botanists objected to the identification of plant and animal motions. How are we to judge whether or not a creature possesses feelings if not by comparing how the creature moves in a given situation with the way we ourselves would move in its place? In plants, "everything seems purely mechanical to us. Their life seems to us less a life than a mere duration. We grow a plant or we destroy it without feeling anything like what we feel when we care for an animal or kill it. We watch the birth and growth of the plant, its blossoming and bearing of fruit, as we watch the hands of the clock move imperceptibly around the dial."[31] This first argument is dubious, of course, because it reflects a kind of anthropomorphism. The second, on the other hand, is conclusive. It has recently been invoked, rightly, to explain the fact that zoological taxonomy was modeled on botanical taxonomy: "Classification requires precision in describing the characters. Precise description requires prolonged observation at leisure. Now, among living things, plants are the ones that are immobile and passive. A wild plant is a plant that was not grown deliberately, not a plant that flees the observer. By contrast, undomesticated animals react to the approach of other animals; that is to say, they try to keep their distance or flee, in keeping with the vital imperative of self-preservation. For a man, a wild animal is not merely an undomesticated outlaw, but a powerful aggressor."[32]

Was this not the reason why the mechanists refused to countenance the existence of plant actions? Man is afraid of wild animals because their actions are unpredictable. But plant motions are easy to study. They are scarcely perceptible, limited in scope, and predictable, and therefore do not disturb the contemplative attitude. These characteristics make it seem as though the cause of the observed motion lies outside the plant. Thus *Anagallis arvensis*, which closes up its leaves at the approach of rain, was known as the "poor man's barometer."[33] This sobriquet sums up the position of the phytodynamicists rather well. The leaves fold as a harbinger of rain. Conversely, when the air is moist, it can be predicted that the leaves will close. Thus the cause of the motion is apparently moisture in the atmosphere rather than instinct. Accordingly the idea that external causes explain the motions of plants gained currency. "The climate governs the lifetime of plants, the time of germination, of foliation, of blossoming, of vigils, of maturity, and of exfoliation."[34] It also governs the turning of the leaves and the direction of growth.

Explanations of plant motion commonly referred to the action of the sun. There were two ways of doing this: either the sun's heat or its light could be adduced as the cause of any given phenomenon. Hill favored light, while Bonnet and Duhamel du Monceau held out resolutely for heat. Hill used the system proposed earlier by Hartley to explain animal motions, applying it to plants. The sun's rays striking a body excite vibrations in the particles of which it is composed. "The change produced in the position of the leaves of plants by light is the result of motion occasioned by its rays among their fibres; to excite this motion, the light must touch these fibres; and where light touches, it adheres, and becomes immediately extinguished."[35] Thus the motion-causing impetus must be constantly renewed. The intensity of the light and the nature of the fiber come into the picture when Hill goes on to explain the various positions the leaves may take. Light of high intensity causes the fibers to become taut; darkness causes them to relax. The amplitude of the motion also depends on the arrangement, consistency, and size of the fibers. If they are long and supple, the amplitude is large. If they are short, compact, and rigid, the motion will barely be perceptible. Not surprisingly, early experiments confirmed Hill's theory. Hill emphasized the role of light by relating variations in its intensity to the opening and closing of the leaves. Once one can obtain a given effect at will, is it not clear that its cause is known? The leaves could be made to move merely by changing the plant's exposure to light. A second series of experiments was then carried out to show that the motion was not due to heat or humidity. The plant was simply placed in a greenhouse that was kept at a constant temperature or kept moist by abundant watering. The opening of the leaves was not disturbed. From this it followed that "two of the four natural agents, heat and moisture, are therefore excluded from any share in this effect. The air is too universal, and its changes too much depend on these, to be admitted in the research. The attention therefore falls on light alone."[36]

Bonnet and Duhamel du Monceau, on the other hand, maintained that the cause of plant motion was the heat of the sun. This necessitated a choice, however, because the heat could act in two different ways, directly or indirectly. It could act directly because the heat is in contact with the outer parts of the plant, especially the leaves. "It suffices to observe that plants have fibers that contract when wet and others that contract when dry."[37] Unlike Dodart, Bonnet carried out a number of experiments, some of which were designed to confirm his theory. The leaves could be caused to move by manipulating the source of heat (a lighted candle, an oven) or moisture (a sponge soaked in water). Other experiments were designed to eliminate light and atmospheric humidity as causes of the observed motion. The motions were the same in dark or wet environments. Duhamel du Monceau, on the other hand, argued that the heat acts indirectly. The heating of the air contributes to the evaporation of the sap. The tendency of the roots to grow down into the earth can then be explained as a result of "the weight of the nutritive juice inside them; and the stems [tend to grow] heavenward on account of the same juice, which, after being treated by the plant, is reduced to vapor and rises in the stems, causing them to grow vertically because of their lightness."[38] Unlike La Hire, Duhamel du Monceau carried out a series of experiments designed to confirm this theory. The first object was to show that the direction in which roots and stems grow is not determined by the alternation of dampness and dryness or light and dark. As in the experiments described earlier, the procedure used to do this was to place the plant in a homogeneous environment so as to eliminate the effects of one of the two factors. To eliminate the effects of dryness, for example, seeds were placed between two moist sponges. To eliminate the effects of light, seeds were placed in a dark room or inside cases filled with earth. The result was always the same: the rootlet curved so as to descend into the earth, and the plumule rose perpendicular to the soil. Similar experiments were carried out on the "sensitive plant" (*Mimosa pudica*), which was placed in a trunk. "Although I managed, in this way, to keep the plant in perfect darkness, it nevertheless opened in the morning and closed at night, just as in the experiment of M. Mairan; this fact assuredly has nothing whatever to do with light."[39] Far less carefully conducted was the other experiment that Duhamel du Monceau carried out in order to verify the theory he had favored from the outset. A seed was placed upside down in a flow of vapor arranged so that it "would push the seed's radicle and plumule into a situation the reverse of the ordinary one, that is, with the radicle above and the plumule below." Duhamel du Monceau reports, however, that "accidents, not worth describing in detail, disturbed this experiment."[40]

By and large the explanations put forward by Hill, Bonnet, and Duhamel du Monceau, though different, are all based on the same principle, namely, that the sun, by conveying light and heat to plants, is the cause of their motions. Hence light and heat are responsible for the opening of the flower, the movement of the leaves, and the ascent of the stem; darkness and cold have the opposite effects. All three approaches are alike in that they move from a narrow initial object of study to a general explanation of motion. Duhamel du Monceau began by investigating the fact that stems grow up and roots grow down, for example. From there he went on to study the sensitive plant and concluded by explaining the opening and closing of flowers. Bonnet first studied the turning of the leaves, then explained heliotropism, and finally the growth of stem and roots. The "sleep of plants," an expression first proposed by Linnaeus[41] and later taken up by Hill, clearly had only a metaphorical significance, however.

It was also in a metaphorical sense that Duhamel du Monceau used the word muscle to denote the fibers of the wood: "The word 'muscle' may be used to refer to a tissue of fibers arranged in such a way that, by contracting, they cause a part to act in a specific way; in a word, these are the muscles of vegetables. However, it must also be admitted that they differ greatly from the muscles of animals."[42] Duhamel was discussing the motion of capsules of fruits and pods of leguminous plants. The muscular system of animals served as a model, but interest centered on the phenomenon of contraction rather than on the complex muscular mechanism. Thus the only resemblance between muscular motion and the motion involved in the opening of the capsules was that fibers under tension played a role in both animal and vegetable. The mechanism associated with this tension in plants was different, however, from that associated with muscular motion. In plants all that happens is that the fruit separates so as to allow the seeds to escape. The next step in the comparison is to contrast the structure of the plant with that of the animal: "The muscle fibers of plants . . . do not form large masses of closely packed fibers but are rather gathered together in small bundles which diverge from one another, and between which are found large masses of cellular tissue." Still other differences are associated with the different mechanisms involved. Animal fibers and vegetable fibers work on different principles. "The contraction of the muscle fibers of animals appears to depend on the juice they contain . . . whereas vegetable fibers contract when they dry out and therefore shrink in every dimension."[43]

Up to the beginning of the nineteenth century, the study of plant motions continued to revolve around the problems outlined here as they appeared in the mid-eighteenth century. Interest centered on the movements of the flowers and leaves and on the direction of growth of stems and roots. The motions of sensitive plants were also studied, including both accidental motions and motions provoked in some way. There were two aspects of phytodynamic research. First the theories of Hill, Bonnet, and Duhamel du Monceau were reexamined. Further detail was added, and corrections were made on points of some importance. In regard to the opening and closing of flowers and leaves, one possibility was to follow Bonnet in arguing that these motions were due to heat. This is what Sénebier did. Unlike Bonnet, however, Sénebier recorded the temperature: "I found that these flowers [crocus, tulip] began to blossom, in the open air as well as in the dark, when the heat affecting them caused the thermometer to rise to eight degrees above zero."[44] Another possibility was to follow Hill in arguing that these motions were due to light. This is what Auguste-Pyrame de Candolle did. His experiments relied on an apparatus far superior to what had been used earlier. Instead of moving the plants, de Candolle placed them in a cellar and exposed them to an artificial light. In this way he managed to reverse their cycles. When *Mimosa leucophala* and *Mimosa pudica* were kept in darkness during the day and illuminated during the hours of darkness, they eventually came to shut their leaves in the morning and to open them at night. Thus de Candolle proved that "light has a very marked effect on the daily movements of the leaves of a number of plants."[45]

Research also progressed in regard to the direction of growth of the stems and roots. No satisfactory explanation of this phenomenon had been given. The first objective was to prove that neither heat nor light was the cause. To do this, John Hunter followed Duhamel du Monceau's procedure, but, rather than put the seeds in a tube filled with earth, he placed them in a tub covered with a trellis. The tub was inverted and suspended three feet above the earth. The bottom of the tub was covered with wet straw to counter the action of the heat, while mirrors were placed on the ground to reflect light into the tub or onto the surface of the compost. "I found that all the seedlings grew up toward the bottom of the tub, and in those where the bud had been placed at the bottom, the young shoots had turned around in order to grow upwards."[46] Knight revived a hypothesis that had been put forward by Duhamel du Monceau. The latter had noticed that the root encases itself in the water-soaked earth: "Another force holds them fast and keeps them from separating from the moisture of the earth, namely, contact with the parts of the water and the adherence of one part to another; for there is no doubt that the moisture of the earth and the sap of the roots constitute a single continuous body, which is subject, like all bodies, to the laws of gravity."[47] But Knight, unlike his predecessor Duhamel, tried to verify this hypothesis.

If gravitation caused the roots to grow downward and the germ to grow upward, it could produce this effect only if the seed remained in a fixed position relative to the attractive force exerted by the earth. This suggested to Knight the following idea: "I imagined that the operation [of gravity] would become suspended by constant and rapid change of the position of the germinating seed, and that it might be counteracted by the agency of centrifugal force."[48] Hunter is probably the first person to have placed a seed in a rotating tube. The speed of rotation, however, was far too slow to overcome the effects of gravity on the seed. Knight used a rapidly rotating wheel (150 revolutions per minute) and was thereby able to counteract the effects of gravity and test his hypothesis.

The other theme of research into plant motion involved motions that were caused either deliberately or by accident, the movements the sensitive plant makes when touched. Many botanists took a view of this class of motion similar to Duhamel du Monceau's view of the mechanism responsible for the opening of the capsules in leguminous plants. The mechanism of animal motion served as a model.

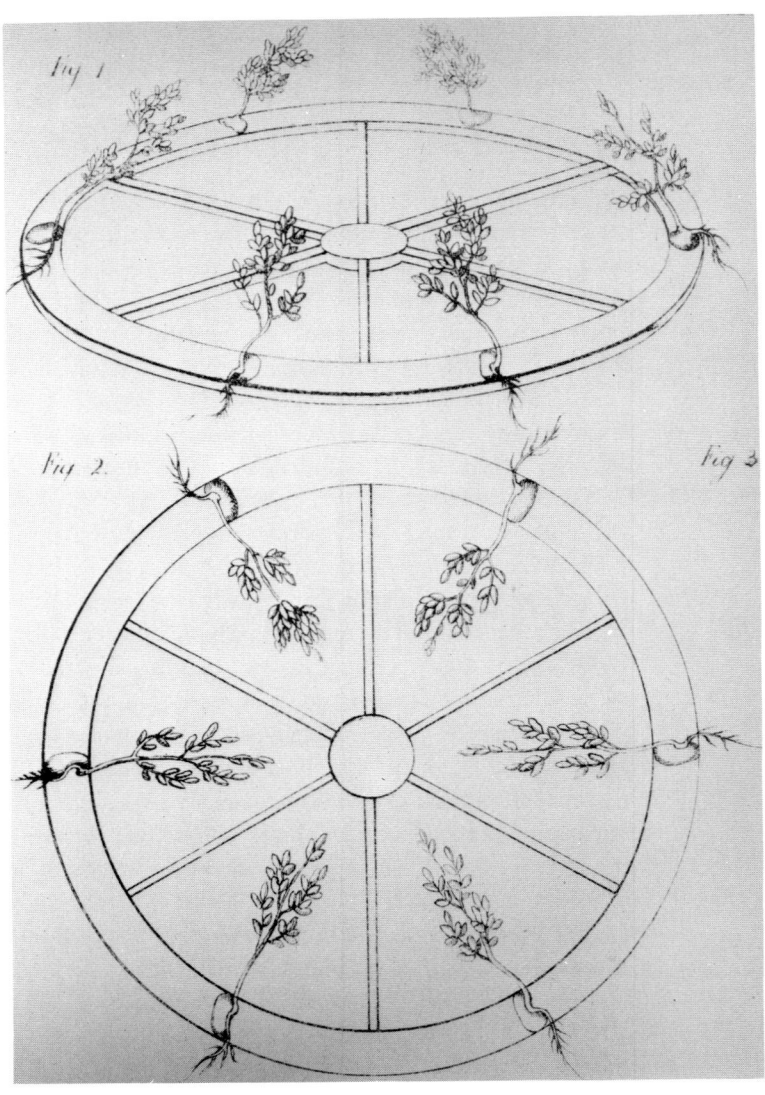

Humphry Davy, illustrations from *Elements of Agricultural Chemistry* (London, 1813). Photo from the French edition (Paris, 1820) courtesy of the Muséum national d'histoire naturelle, Paris.

MOVEMENT

Attention centered not so much on the muscular apparatus per se as on the contraction of the fibers. The tension of the fibers was the point in which sensitive plants and animals were thought to resemble one another. Whether this tension works in the same way in sensitive plants as it does in animals was the next question to be asked. "The movements of animals are, I believe, universally attributed to a simple contraction of the fibers known as muscles. Is the movement of the sensitive plant also accomplished by a simple contraction of the fibers in a manner similar to that of animals?"[49] Clearly, in order to answer this question, the effects due specifically to irritability had to be compared with the motions of sensitive plants. There were only two possibilities: if the effects observed were similar, botanists would have to recognize the existence of a "principle of irritability" in this plant; if, on the contrary, differences were noted, they would be left with the task of figuring out how the sensitive plant moves. To approach the study of motion in sensitive plants thus, it was necessary first to define the characteristics of irritability in a precise way. What did this entail, if not to determine similarities and differences in the effects due to stimulating the various parts of animals and plants?

One possibility was to characterize the irritability of animal fiber, as Whytt did, as a convulsive motion. Whatever the stimulus applied, contractions alternate with relaxations. The only motion of the sensitive plant, however, is the folding of its leaves, and this occurs whenever they are touched. Thus, "there are no alternate contractions and relaxations, as in the muscle fibers of animals; there is no indication of feeling or of being particularly affected by stimulating substances; but everything is accomplished by mere contact or by a mechanical impulse."[50] Another possibility was to take repeatability of the effect as the criterion of irritability, as Lamarck did. Contraction occurs whenever a stimulus is applied. "The effect [of this stimulus] consists in an instantaneous contraction of every irritable part of the animal that is in contact with a foreign body, a contraction which ends when its cause is removed and which, after the part has relaxed, repeats itself each time the irritation is renewed by further contact. Nothing of the kind has ever been observed in any part of vegetables."[51] Finally Lindsay undertook a series of experiments similar to those that had disclosed the irritability of animal fibers. Dropping water on the stem did not result in any motion. A branch could be cut or the stem of a leaf pierced through with a needle without causing the leaves to fold.[52] It was therefore clear that sensitive plants do not react in the same way as irritated muscles and that the motions of such plants are different from motions due to irritability.

This meant that it still remained to describe the mechanism responsible for the motion of the sensitive plant along with the structure underlying that mechanism. A comparison was made with the mechanism and structure of the muscle. The point of this exercise was to specify the differences in structure and function between the plant and animal mechanisms. As for structural differences, "the joints of animals are composed of two main parts, the bones and the hinge. The joints of sensitive plants are also composed of two main parts, one made up of a bundle of longitudinal fibers and the other of a substance surrounding these, which I have called the cellular substance. This cellular substance is the part in which the principle and power of movement reside, just as in animals."[53] Thus it is not fiber that is the agent of motion, because fiber plays a supporting role, like bone. Further, as to the contrast between the mechanisms, the first thing to note is that animal motion results from the contraction of muscle, which requires the introduction of a nervous fluid. In plants, the leaflet opens because of tension in the cellular substance. In other words, an elastic fluid collects in the joints and expands the parts. A touch causes these fluids to dissipate and the cellular substance to collapse. This phenomenon therefore cannot be compared with the movement of the fibers in animals. "When I touch the extended branches of the sensitive plant [*Mimosa pudica*], instead of a *contraction*, what I observe in the joints of the shaken branches and stems is an immediate relaxation, which allows the branches and leaf stems to collapse so that the leaflets themselves fall one on top of the other."[54] Lindsay proposed a similar explanation: "The flexing of the joints of sensitive plants is not caused, as in animals, by the contraction of any fibrous part; rather, it is clear . . . that the flexing of these joints is, mainly, nothing other than a surrender or inclination in the direction of the least resistance."[55] Finally, Broussonet applied this explanation to the motions of *Dionea muscipula:* "The glandules, which can be seen in the midst of each leaf of Dionea, are no sooner touched by some insect than the leaf folds back on itself and immediately grabs hold of the animal: the touch seems to cause a release of the fluid that held the leaf open by filling up its vessels."[56]

Proponents of plant mechanism also examined the proofs that were put forward in support of the plant action thesis. The object was not to question the observations and experiments themselves, but rather to question their interpretation. For one thing, it may well be that opium or electrical discharges suppress plant motions, but it is equally possible to argue that "the means used to suppress the irritability of animals are means that disrupt the organization of the plant. They are frequently lacerated by a strong electric current and always dried out by a bright light; opium alters their surface by covering them with a gummy-resinous substance."[57] If the poisons that affect animal nervous systems are also mortal for vegetables, it does not on that account follow that vegetables have a nervous system. Then too the motions of climbing plants that grasp a thin pole and twist around it no matter where they are placed do not indicate the existence of any sort of instinct. "Climbing plants and tendrils grow in all directions, one after another, and consequently cannot fail to come into contact with any objects placed within their reach."[58] Finally the movement of the roots, far from indicating the existence of any instinct, is related to the rapid growth and development of the roots themselves, which are in contact with the nutritive soil. This explains why Knight pronounced himself "wholly unable to trace the existence of anything like sensation or intellect in plants."[59]

THE QUESTION OF THE DISTINCTIVE CHARACTER OF VEGETALITY

When the physiologists had completed their investigation of vegetable motion, one question remained unanswered: Does vegetality have a distinctive character? Those who, like Lamarck, explained the motions of plants in terms of the laws of mechanics obviously had a ready-made answer: "Vegetables are non-irritable living bodies, incapable by themselves of causing any of their parts to contract instantaneously and repeatedly, and without any faculty to act or to move from one place to another."[60] It will be objected, perhaps, that, well before Lamarck, Haller and Scopoli had already said the same thing. For Haller, in fact, irritability was the feature that distinguishes animal fiber from vegetable fiber. In 1743 he "announced that every irritated animal fiber contracts, and that this character distinguishes it from vegetable fiber."[61] For Scopoli, "a plant is a natural, extensible, non-irritable organic body, which grows by intussusception and which is therefore a vegetating living thing, but insensitive and unable to move at will."[62] In fact, neither Haller nor Scopoli really distinguished between plants and animals in an unambiguous way. Haller contradicted himself in later work: "The example of polyps and other insects that have neither a brain nor nerves and yet can easily be irritated proves the great potency of irritability in the animal fiber; furthermore, the same thing is proved by the similar reactions of many different plants, which open and close their flowers and leaves depending on the degree of heat or cold, and which in some cases seem to act just as quickly as animals, if not even more quickly."[63] As for Scopoli, it is clear that he treated irritability not as the distinctive character of vegetality, but rather as an additional character not essential to the definition of plant life.

The question is embarrassing for anyone who attributes to plant life a property of irritability, to say nothing of some sort of sensibility or instinct. The reason for embarrassment is that it must be acknowledged that vegetality has in this respect no distinctive character. Nevertheless, there is no reason on that account to confound plants with animals. Thus, the question, What is the nature of the difference between these two classes of organized beings? supplanted the insoluble question as to the distinctive character of vegetality. Two requirements had to be met: plants had to be distinguished from animals, but this had to be done in such a way as not to create too sharp a breach between the two kingdoms. For this, it was enough to assume that all animal functions are also found in plants, and that the only differences between the two kingdoms are structural. Positing a principle of functional invariability made it possible to assert that no sharp dividing line separates one kingdom from the other. Insisting, further, on the variety of the anatomical structures whereby those functions are accomplished made it possible to determine the differences between plants and animals. Of course, it is true that the uniformity and simplicity of plant organization made it impossible for investigators to follow in detail the modifications affecting the anatomical substrate of every function, sensitivity in particular. It was easy enough to get around this difficulty, however. In fact, rather than posing an obstacle to proof of the theory, the difficulty became an argument in its favor. "The nerves of plants are probably as different from the nerves of animals as their structures are different."[64] Doubtless nutrition provides the best example of a gradual degradation or simplification of structure that does not obscure or eliminate the constancy of function. "At the lower end of the scale, we see organized bodies whose fluids are simply shifted back and forth from bottom to top and from top to bottom. A little higher up, we find other bodies whose fluids move in different directions. If we climb still higher, we discover a rudimentary circulation, the apparatus of which is reduced, however, to one or two large vessels. This apparatus becomes increasingly complex as we move up still higher on the scale."[65]

The idea of a variation affecting only the anatomical organs was still only theoretical, however, and full experimental verification was a long way off. Indeed, a disparity very quickly appeared between the notion that, as one moves down the chain of being, the functions progressively become simplified or reduced and the results of actual observation of animal and plant structures. Hence a good many naturalists continued to believe in the existence of distinctive characters. Among those mentioned were digestion (Boerhaave), reproduction, irritability, locomotion, and even the form of the animal. For those who accepted this and who saw "no reason why we may not include Vegetables in the class of Animals," since "gradation can only be admitted in beings of the same nature in some respects, but not in those which entirely differ from one another,"[66] and who acknowledged, further, that it was impossible to demonstrate this gradation experimentally, there was only one possible course, namely, to refute what other botanists proposed as distinctive characters by suggesting counterexamples. Among others, Bonnet and Robinet proceeded in this way. Some botanists had thought it possible to distinguish between animals and plants by virtue of their shape, but "animal life is so prodigiously varied in the multitude of its shapes, that this variety alone is itself sufficient proof that no exterior shape is peculiar to the animal kingdom and none is excluded by it."[67] Another tack was to distinguish between plants and animals by virtue of their mode of nutrition. Animals feed themselves through internal roots, plants through external roots. Bonnet believed this choice of distinctive character to be a poor one: "What makes it ambiguous is the existence of creatures that are truly animal and yet feed themselves, as do various plants, using their whole constitution."[68] As an example he cited the tapeworm, whose body is covered with orifices or suckers. Nor can the faculty of locomotion serve as a distinctive character of animality, because it is lacking in oysters and other shellfish. Ultimately those botanists who believed in the existence of animal actions in plants succeeded in demonstrating that there is no distinctive character of vegetality. The proponents of plant mechanism, on the other hand, distinguished plants by a negative criterion, the absence of irritability.

ETHICS AND AESTHETICS OF VEGETALISM

Histories of botany generally take the view that the conflict between the proponents of plant mechanism and the believers in plant actions stemmed from a principle whose main purpose was to satisfy certain theological requirements. Alphonse de Candolle, among others, emphasized this point: "Some philosophers, in the main guided by religious ideas, did indeed try to prove that vegetables are endowed with sensitivity, that they are conscious, for example of their existences and perhaps aware of sensations. Others, starting from the same ideas and from the fact that vegetables have no organs of motion, believed that it would be contrary to the order of the universe and to the goodness of one of the attributes of divinity that beings should be endowed with the faculty of sensing evil without being able to flee from it, and with the faculty of desiring the good without having the means to attain it."[69] On the one hand, we are told, were those eager to exalt the power of the Creator; because of this ambition, presumably, they endowed plants with the faculty of sensitivity. On the other hand, ostensibly, were those eager to exonerate God of the charge that he had left defenseless creatures to suffer helplessly; this, presumably, was their motive for suggesting a mechanistic explanation. Conceived in this way, the subordination of the perception of phytodynamic phenomena to themes drawn from Christian theology is significant in two ways. First it served, in hindsight, as justification. Once plant motions had been explained by the laws of physics, an obstacle worthy of the stakes at issue had to be found in the form of epistemological necessity. In other words religious ideas had to lose their power to perplex so that belief in physics could replace the physics of believers. Second this justification made it possible to paper over a remaining problem. On the pretext either of exonerating God or exalting his wisdom, the real object was nothing less than to exonerate man from the charge of destroying plants.

But what are these sharply contrasting religious ideas that are supposed to have structured the various representations of the phenomenon of plant motion? Some botanists favored the Cartesian line of argument, at any rate as it was spelled out by Boerhaave or Hoffmann. Specifically they believed that man possesses a soul, that animals are merely living machines, and that plants are mere machines. Anyone accepting this line of argument was obliged to deny the existence of sensitivity in vegetables. "It seems contrary to the general process of nature that creatures which can neither defend themselves against evil nor flee from it are endowed with the faculty of feeling it; furthermore, it has been observed that all the differences between the two kingdoms can be deduced from this one difference alone, to wit, that animals possess sensitivity, that is to say, awareness of their existence, while vegetables are deprived of it."[70] Other botanists favored the line of argument underlying the ideas of natural economy. They therefore believed in the existence of sensitivity in vegetables. "Is it imaginable that such profusion of life can exist without the slightest sensation? It is far preferable to assume that vegetables share this faculty with animals, and that our great Creator has apportioned the good among all living creatures."[71] In fact, religion suggests not so much arguments as solutions to a unique problem. In order for that problem to emerge, the factors have to be put back in the right order, that is, in the present case, reversed. Whether the existence of sensitivity in vegetables is affirmed or denied, the same requirement must be met, namely, the requirement to justify the use or destruction of plants by animals and men. For physiologists, then, the references to the creator thus served to open two paths to the same goal: to do away with the problem of suffering in plants in order to exonerate man of the sin of injustice in making use of them.

In one respect, to prove the Creator's goodness is to justify man himself. If God did not allow plants to feel pain, then man has the right to destroy them: "Would it seem worthy of the wisdom of the Creator that vegetables, which are prey to animals and exposed to the elements, should possess sensation and volition?"[72] The ethical position upon which this denial of sensation is based is immediately apparent. Peschier is doing for plants what Descartes did for animals, devaluing them in order to allow man to use them. The scope and meaning of this solution to the problem are the same as in the theory of animal-machines. The existence of a soul—of reason—implies knowledge of good and evil. Thus man can either respect moral laws or transgress them. It follows that pleasure is a reward and pain a punishment. It is easy to understand why the existence of a soul in animals would have been a sign of injustice. In giving animals a soul, God would have given them the power of awareness without the means and therefore without the obligation to think. Animals lack language, the outward sign of thought. They are therefore unable either to respect or to violate a moral order of which they are ignorant. The pleasure they experience is not a reward, and the pain they feel is not a punishment; thus there is no injustice in inflicting pain on them. To deny the existence of a soul in animals is not the same as to deny them sensitivity, however. This raises a question: Why were some botanists not content to treat plants in the same way that animals had been treated earlier? Why did plants have to be denied not only a soul but also a sensitive faculty? The answer is that animals were compared to man but plants were compared only to animals. In animals, sensitivity plays a part in self-preservation, for the animal seeks what it needs and flees what may do it harm, protects and defends itself. Thus the existence of sensitivity in plants would be the sign of an injustice. The plant would have the faculty of feeling without the means to act, since it lacks a locomotive principle and prehensile and defensive organs.

It would seem that anyone who raised the plant to the rank of a sentient being and looked upon its movements as more than merely mechanical should have been obliged to renounce vegetalism. From sentimentality to impiety was only a short step. It was indeed presuming a great deal to think that God would not allow apparently defenseless creatures to be made to suffer. "If we contemplate only the evil that may result from each of nature's institutions without considering the good, it may seem irreconcilable with the idea of a merciful and charitable Creator . . . but we must not take the view that the laws of nature accord with our biased view of their effects. We must consider the general import of these laws and the designs they serve before we declare them unjust and oppressive."[73] The suffering of plants, like that of animals, should therefore not arouse compassion. Without the war of all against all, the species would proliferate and the natural balance would soon be upset. From this standpoint, the death that lies in wait for all living things, and hence the suffering that they experience, is not the sign of a fall from grace or an injustice, but a tribute to the divine order. Man, like animals, is thus exonerated of the sin of injustice in killing plants.

If plants suffer, moreover, they can also inflict suffering on the creatures around them. They are capable not only of defending themselves but also of causing death. Their means of defense and attack are of course different from those of animals. The plant's means of defense are nevertheless entirely appropriate to its type of existence and quite as deadly as the animal's weapons. In the first place, plants, like animals, are equipped with defensive weapons: "just as claws, nails, horns, and beaks serve animals for defense, so, too, are plants provided with the same resources."[74] In addition, they have offensive weapons: are not the actions of certain plants more expressive and clever than those of many lower animals? Study of the Venus fly trap, for example, had shown that "nature was determined to provide for its nourishment by forming the upper portion of its leaves in such a way as to enclose an instrument contrived so as to grasp whatever food should offer itself, and by placing in its center a lure to attract the insects destined to feed the plant."[75] Finally, because plants experience pain, they also feel pleasure. "May not the exercise of their vital functions be attended with some degree of sensation, however low, and some consequent share of happiness? Such a proposition accords with all the best ideas we can form of the Divine Creator."[76]

To be sure, the idea of a kind of happiness in plants is a thought that no doubt gave narcissistic pleasure to those who believed in the existence of sensitivity in plants. "I like to persuade myself that the flowers that adorn our fields and gardens with ever fresh splendor, the fruit trees whose

John Ellis, table II from *De dionaea muscipula planta irritabili nuper detecta ad Car. a Linné epistola* (Erlangen, 1771). Photo courtesy of the Muséum national d'histoire naturelle, Paris.

fruits have such pleasant effects on the eyes and palate, and the majestic trees that fill vast forests untouched by time, are so many sentient beings, which in their own way taste of life's pleasures."[77] The believers in plant mechanism shared the taste for spectacle, however. Linnaeus pointed out how interesting it would be to put together a floral clock: "A *Floral Clock* should be assembled for each different climate, governed by the awakening of the different plants, so that every man might know the correct time of day, even if he has no watch or the sun is hidden."[78] Of a practical order, this justification seems to have hidden a preoccupation of quite another kind, aesthetic in nature. Linnaeus was in fact employing a rhetorical device that had been commonplace in the first half of the seventeenth century. Originally, the flower was an image of time. Later, the relation between time and the flower was reversed, emphasizing the latter: time is a flower. Linnaeus shifted the burden of the metaphor to this new center, borrowing an image from a concrete object: time equals the clock. Finally, the flower was itself represented metaphorically by Flora (flower equals Flora). The result was the floral clock. The object itself might be classed, along with the garden motifs, in the category of "barrochus rupestris," as Eugenio Ors has suggested. Ors notes that "the essence of the baroque always contains something rural, something of the peasant."[79] The floral clock indeed meets all the criteria of the baroque. For one thing, time is epitomized in sensible form in the work. For another, the flower is, as usual, an image of time and its inexorable flight. In a curious reversal, however, the flower is also the device by which time is measured and broken down into its components. The work, an artificial object constructed of natural components, the flowers, is clearly perishable. The Linnaean motif thus denies itself, for it is transitory and fleeting. Finally, the decorative takes precedence over the functional value: "A Floral Clock should contain flowers that display their full magnificence at different hours of the day."[80]

To sum up, then, the bone of contention between the proponents of the plant action theory and the supporters of plant mechanism is clear. By choosing irritability, sensitivity, or instinct as models, the plant action theorists came to the view that plants possessed a property and faculties related to that property that seemed characteristic of animals. When similarities of cause and effect are looked for from the outset, it is inevitable that observational techniques will be less than rigorous. The systematic emphasis given to apparent similarities of motion leads to a claim that the underlying physiological mechanisms are identical.

The rivals of the action theorists, in seeking to establish the validity of the mechanistic theory, chose to base their comparison on the purposes to which animal actions are directed. They therefore had to begin by making a choice: either the mechanisms of plant motion are similar to those responsible for animal motion, or they are different. To decide one way or the other, the mechanists were forced to undertake comparative study of animal and plant motions. In the course of this work, differences were noted, which led to a choice in favor of the second of the two alternatives. The comparative work was based on animal physiology, which guided both the comparison between plant and animal motions and the construction of mechanisms to account for motion in plants. Ultimately the mechanisms of plants were said to be different from those at work in animals.

Furthermore, although dal Covolo and van Marum did do experiments, they always interpreted the results in such a way as to reinforce the prior structural description, which was based on observation. Lindsay and Lamarck, on the other hand, did not confine their attention to apparent resemblances but rather focused on the different effects obtained by applying stimuli to various parts of plants and animals. Dal Covolo and van Marum quickly came up with a representation of plant life that was at once complex and enigmatic. Plants were said to be irritable, sensitive, and even endowed with a kind of instinct. Yet it was impossible to pinpoint the location of their nervous systems. Lindsay and Lamarck, on the other hand, invoked the laws of physics to explain the reactions and motions of plants.

Once it is shown that plants perform the same functions as animals, it becomes impossible to draw a sharp dividing line between the two kingdoms. Accordingly no distinctive character of vegetality can possibly exist: "Withdraw from the notion of Cat or Rosebush all the properties pertaining to the species, the genus, and the class, and retain only the most general characteristics of the animal and the plant, and no sign will remain to enable you to truly distinguish between the Cat and the Rosebush."[81] By contrast, the plant mechanists were able to suggest irritability as the distinctive character of animal life.

Finally, religious ideas did not by any means govern the construction of knowledge, but, as always, themes derived from those ideas could be used to interpret the concepts of plant physiology. Those opposed to the notion that plants are in some way sensitive emphasized the fundamental goodness of nature, or, again, of God, who would not allow defenseless creatures to be made to suffer. Those who believed in the existence of sensitivity in plants invoked a God wholly preoccupied with the balance of nature, for whom suffering had its raison d'être. In actuality, the point in both cases was to exonerate man of guilt as a destroyer of plant life. The theological concepts underlying these two different lines of argument could both be made to accord with the naturalists' view of plant motion.

Conclusion

Rather than accept the contrast that has been drawn without justification between the proponents of analogy and the adepts of observation and experimentation, I have tried to look at the statements made by members of both camps, analyzing the way in which the focus moved from analogy to observation to experimentation. Rather than separate what is supposed to come under the head of positive science from what is allegedly without scientific interest, I have tried to take seriously notions that are taxed with being false. Thus I have taken the view that the rules of botanical discourse led to the formulation of what Michel Foucault would call "disciplined errors."[1] In order to avoid the pitfalls of psychologism, I have tried to describe the various positions that a given individual thinker *might* have taken by examining the analogies, models, and comparisons that were available and the manipulations and experiments that were possible. Finally I have subjected what appear to be causal processes to scrutiny and have detected not continuities but divergences and differences. In constructing each series of problems, I showed how the problems involved were reformulated and modified, sometimes in important ways, sometimes in ways not so important. I have looked at the ways in which lines of investigation twist and turn and in some cases even come to a halt.

Reasons have been given for the epistemological priority enjoyed by plant taxonomy during the eighteenth century. There were first of all economic reasons. It was useful to have an orderly inventory of the resources available in the plant kingdom. There were also reasons of a biological order. Careful description of the characters required close observation, and it was far easier to observe plants than animals. Finally the plant, unlike the animal, is defined by its external features and by the simplicity of its structure. Since the external parts can all be seen, taxonomical knowledge is easily acquired. Vegetality, though, was in a sense the mirror image of taxonomy, which explains why, in physiology, epistemological priority belonged to the animal kingdom. Plant physiology in fact followed in the wake of animal physiology and modeled its objects on those of its predecessor. Plant physiology was slow to develop because the apparent simplicity of the plant made it impossible to say what role was played by each of its parts. By contrast, it was easy to talk about the uses of the parts of animals, because technological artifacts were more like animals than they were like plants. It was therefore possible to decipher the structure of animals by comparing their internal organs, which are easily visible, to mechanical devices. Knowledge of animal life consequently outstripped knowledge of plant life. But why did plant physiology have to model its objects on the objects of animal physiology? Once interest comes to center on the vital phenomena that plants and animals share in common, it can be assumed that plants perform functions like those performed by animals. As soon as functions entered into the discussion, an abundance of models was available in the animal kingdom. Thus higher forms of life could be used to shed light on lower forms. A statement fell within the purview of plant physiology only if it pertained, as we have seen, to a specific set of concerns, the functions of the plant—that is, the mechanics of fluid motion within the plant body, the mechanisms of reproduction, and the motions of the plant.

CONCLUSION

At the beginning of the nineteenth century, however, the "discursive practice" I have been describing broke down. In its place a scientific practice was elaborated in accordance with different rules and related in a new way to animal physiology. The plant in effect became the laboratory of cellular theory. The tables were turned; from this point on, the lower forms of life were to shed light on the higher. Of course the object of study was now quite different, interest centering not on organs and functions but on the intimate structure of the plant, its fundamental composition. Let us briefly recall the reasons for the preeminence of the research in plant anatomy carried out by Konrad Sprengel, Brisseau de Mirbel, and Henri Dutrochet. In the first place this work was important because its object was the cell. Because the cell wall is far more clearly defined in plants than in animals, it was possible to recognize the cellular structure of plant tissue. Second, there were technical considerations; using heat or acids, it was easy to isolate the cells of plants, whereas study of animal tissues required much more delicate methods. Even more important, cellular research in animals was halted because of the impact of Bichat's fundamental work on general anatomy. The question became the classification of tissue rather than the nature of its parts. Meanwhile, Dutrochet studied the workings of the cell,[2] and Robert Brown discovered the nucleus. When Schleiden and Schwann extended cytology to the animal kingdom, cellular theory at last became established on a firm footing. The cell was shown to be the unique building block of both kingdoms, the fundamental component of living organisms.

The attack on reproduction also shifted to new ground. Not floral structure but the microscopic study of fertilization became the center of interest. Using the first achromatic objectives and immersion techniques, Amici in 1823 was able to observe the formation of the pollen tube in phanerogams, the emission of pollen, and its growth inside the tissue. Fertilization was still liable to be interpreted in divergent ways, however. In Schleiden's view, the end of the pollen tube was the first adumbration of the embryo, since it pushed back the membrane of the embryonic sac and lodged itself there. Amici held instead that the egg-cell existed inside the embryonic sac prior to the appearance of the pollen tube and that, further, it was the contact of the end of the tube with the embryonic sac that accomplished fertilization. These assertions were confirmed somewhat later by Hugo von Mohl and Wilhelm Hofmeister. Mention should also be made of the study of cryptogams undertaken by the latter, who formulated the "law of alternation." This blurred the dividing line between phanerogams and cryptogams; Thuret observed the fertilization of algae, and Pringsheim described the penetration of the male part and observed the formation of a membrane that cut off access to the fertilized egg or oedogonium by any other part.

The laws of nutrition in plants were based on the substitution of Lavoisier's system of chemistry for phlogistic chemistry, as well as on the introduction of quantitative methods. The transition to a physicochemical explanation of vegetation took place during the last quarter of the eighteenth century. In 1772 Priestley showed that the toxicity of the air caused by the combustion of a candle was eliminated by the presence of a green plant. In 1793 Lavoisier proved that in darkness the respiration of living things, both plants and animals, decreased the quantity of oxygen and increased the quantity of "carbonic gas." Although Ingenhousz showed that peculiar properties of green leaves come into play only under the influence of light, it was Sénebier who established, between 1782 and 1788, that plant respiration was the inverse of combustion, that the leaves absorb and decompose the carbonic acid in the air, fix the carbon, and eliminate the oxygen. The crucial step was taken in 1804 when Théodore de Saussure, using quantitative measurements of weight, showed that the bulk of the dry matter

CONCLUSION

of plants was composed of carbon, hydrogen, oxygen, carbonic gas, and water, and that a part of the oxygen was given off into the atmosphere. He also determined the quantities and kinds of substances that entered into the organization of the plant. The botanists learned from the chemists.

In 1828 Dutrochet explained plant motions in terms of changes in turgidity resulting from endosmosis. Mechanically the effect of endosmosis is an endosmotic flow toward the more concentrated solution, along with a reverse flow. Mention should also be made of Hugo von Mohl's work on the structure and motions of climbing plants. He not only pointed out the differences between voluble stems and tendrils but also discovered that the contact between the tendril and its support had the effect of a stimulus.

Such was the new landscape at the end of the eighteenth century and the beginning of the nineteenth. What we are witnessing is nothing other than the birth of plant physiology. Doubtless one could sift through the writings of eighteenth-century botanists and separate those notions that fit in with nineteenth-century science from what belongs to an outmoded ideology, but this would be to blur the difference between natural history and phytobiology.

To sum up, I have been obliged to take the view that the classical era of botany was neither an epilogue to the Renaissance nor a prologue to the nineteenth century. I have had to write the history of a discipline, therefore, that cannot be classed as biology, a discipline that comes after Descartes but before Lavoisier and Bichat. In studying vegetality as it was conceived in the eighteenth century, I have attempted to analyze a practice that worked out its objects in accordance with precise rules and have been concerned not with the genesis of meanings, but with the production, the generation of botanical propositions.

CONCLUSION

Appendix

The text that follows is an eighteenth-century translation of Antoine de Jussieu's *Du rapport des plantes avec les animaux tiré de la différence de leurs sexes* (ms. 284, 1721, Bibliothèque du Muséum d'histoire naturelle). The translation is that of Richard Bradley, *A Philosophical Account of the Works of Nature* (London, 1721), pp. 25–32 [here cited from the 1739 edition, where the passage comprises pp. 38–48—trans.].

The Analogy *between* Plants *and* Animals, *drawn from the* Difference *of their* Sexes.

That *Plants* and *Animals* are *analogous*, we may be convinced, if we only consider the *Manner* whereby they receive their *Nourishment*. That sort of Life, which the antient Philosophers observed in Plants, was accounted by them so nearly the same with that in *Animals*, that they did not scruple to call it a *Soul*, but has since been more reasonably termed *Vegetation*.

The Comparison that has been made between the Structure and Use of the *Bark* of *Plants*, with the *Skin* of *Animals*; of the *Tubes* through which the *Sap* is conveyed through the *Trunk* to the *Extremities*, with the *Arteries* and *Veins*; the Resemblance of the *Ramifications* of those *Channels*, which the *Blood-Vessels* and *Lymphaticks*, has given Occasion to Cesalpinus and other illustrious Authors amongst the Moderns, who have studied the *Anatomy* of *Plants*, to establish this System.

We may yet advance this Opinion much farther, if we consider the *Nature* of *Plants*, and how they may be distinguished as *Terrestrials*, and *Aquatics*, and thereby agreeing with the *Animal* Kingdom. We may also compare the Solidity and Duration of *woody* and *vivaceous* Plants, with the Strength and Length of Life in *Quadrupeds*. We may likewise observe the Similitude between the *Capillary* and *Fungous Plants*, and the short Remain of such as are *Animal*, with the Imperfections attributed to *Insects*, and the Shortness of their Lives: And to this we may add another Remark, That among *Plants* there are *two* sorts of *Aquaticks*, which, like Fish, are either distinctly Inhabitants of the Salt or Fresh Waters.

As there are *Amphibious Animals*, so is the *Vegetable* Kingdom also furnished with *Plants* that have Parts which live as well *out*, as *within* the *Waters*.

But as these general Observations, which are founded upon the Structure of the *Organs*, and upon the Mode of *Growth*, have been already discussed by so many Physicians, that there is no room left for Doubt: I shall make it my Business to establish another kind of Agreement betwixt *Plants* and *Animals*, by more particular Observations, drawn chiefly from the *Diversity* of *Sexes*, and from the Conformity and Uses of those Parts which are to be distinguished in them, for the perpetuating of their *Species*.

It appears that the Antients had some Notion of this Distinction of *Sex* among *Plants*, as we find in their Writings that some have given the Quality of *Male*, others of *Female*, to certain *Vegetables*; but the more I have taken Pains to examine into the Reasons they had to establish this *Difference* of *Sexes*, so much the more I find them out of Reason; especially when I discover that they have given the *Feminine*-Character to some *Plants* for the sake of their beautiful *Flowers*, or from the Port or Appearance of the whole *Plant*, as in the *Peony* and some others; or else they had established the *Masculine*-Gender from the Conformation of the *Roots*, *Fruits*, or *Seeds*, as they were nearer the Resemblance of the *Male* Parts of *Generation* in *Animals*, as in the *Orchis*, *Mercury*, *Hemp*, &c.

But the Chief among modern *Botanists*, *Malpighius*, *Grew*, *Ray*, and *Camerarius*, have greatly improved upon the Hints given by *Cesalpinus*, by marking out to us in a particular Manner the *Distinction* of *Sexes*, in the Description of certain Parts which perform those Functions.

I shall not in this Place enter upon the late Proposition of Messieurs *Geoffroy*[1] and others, who in the Description they give us of the *several* Parts of a *Flower*, tell us, that the *Dust* which falls from the *Apices* of *Flowers*, is the *Germ* of the *Plant*, or *Embryon* of it; since this System is subject to the same Difficulties with that of the *Generation* of *Animals*, supposed to be affected by the *Worms* or *Animalcules* in the *Male*-Seed.

Nor do I pretend to assign any Reason why Nature has observed so much Regularity in so many different Figures, as we find in the *Farina* of each respective Kind of *Plant*, with the help of *Microscopes*; since Dr. *Grew*, who was so careful in his Observations of this Nature and, Monsieur *Geoffroy*, who is a diligent Follower of our ingenious Countryman, have neither been able to find them out.[2]

I shall only take upon me to compare the *Exteriors* of *Plants* with *Animals*, as far as it regards their *Sexes*, wherein this *Difference* of *Sex* in *Plants* consists, and the Manner of observing it.

As the *Flower* is that Part of a *Plant* which contains the *Organs* for its *Generation*, it is necessary to determine what Idea we ought to have of it, and not to fall too hastily into the enormous Opinions of some, who believed with *Malpighius*, that their most essential Parts were no more than *Viscera* appointed for separating the excrementitious Juices.[3]

I am of Opinion, that what we ought properly to call the *Flower*, is the Assemblage of little Threads, to which the *Botanists* have given the Name of *Stamina*, and are terminated at their *Tops* by *small Caps* or *Purses* called *Apices* or *Chives*, which generally have double Openings, from whence flies out the fine *Dust* which ripened in them.

The *Stamina*, which I have just now mentioned, are either encompassed by a single or double Furniture, either of *one*, or of *many* Pieces, consisting of *one* or of *various* Colours, which have hitherto been called *Leaves*, but may rather bear the Name of *Petals*, to distinguish them from the common Leaves of *Plants*, which are generally *Green*, as well as those which serve for an outward Coat to the *Petals*, and known by the name of *Calyx* to the Flower.

These *Stamina* encompass, for the most Part, a Body of a different Figure, either single or composed, which is either the *Embryo* of the *Fruit*, or a *Tube* terminating like a Trumpet, to either of which the Name of *Pistillum*, or *Pistil*, is independently given.

This Description I conceive is exact and full enough to give us a quite different *Idea* of the *Flower* than what we have hitherto received, and may oppose the vulgar Opinion, that every *Body* of *various-coloured Leaves* is a *Flower* or *Blossom*; instead of which, the curious Observers of Nature cannot miss the Observance besides (besides the *Petals*) of all the other Parts which we have just now mentioned; and from the instant they behold them, must of necessity perceive their Uses from their Structure and Disposition, but more especially when they have Opportunity of observing their several States and Changes at different Times,

We may conclude then that the *Secret* of *Generation* is neither to be found in the *Root*, *Trunk*, or *Leaves* of a *Plant*, but only in the *Garniture* of those *Organs*, which we have observed the *Flower* is composed of: Since we do not find in any other Part of *Plants* those Organs which so well agree with the Parts of *Genera-*

tion in *Animals*, or are so useful and necessary to perpetuate their *Species*. It seems as if Nature, who has hid from us her Manner of Working in the *Generation* of *Animals*, is more inclined to open that Mystery to us in the *Vegetable* Kingdom, since the Means she makes use of, with regard to the *Generation* of *Plants*, is more open, and may be more easily observed.

In effect, what can more resemble those *Organs* which constitute the *Male*-Sex in *Animals*, than those which characterize the same in *Plants*? What Agreement is in their Functions! Those little *Caps* which make the *Chives* of the *Stamina*, the *Farina* which they inclose, and is so exquisitely prepared as a proper Matter for *fecundating* the *Germ*; and again, those Trumpet-like *Tubes*, which are stiled *Pistils*, situate in the Center of the *Stamina*, for the more easy Reception of the *Dust*, which is the Offspring of the *Apices* or *Chives*; the Springs which open them, and the Manner of their flinging abroad this prolific *Dust*: Do not these sufficiently set forth the beautiful Simplicity that Nature observes in her Works?

If we could find modest Terms to express the Care and Precaution which Nature takes to succeed in her Work of *Generation* in *Animals*, what Agreement and Uniformity should we not find with that she makes use of in the *Generation* of *Plants*? The Usefulness of the *Petals* which encompass the *Apices* to press them towards the *Pistillum*, so that their *Dust* may fall in great Abundance into it; as likewise how necessary they are to protect those tender Parts from the Injury of the Wind, may still afford us fresh Matter of Admiration.

It is easy to judge, from the Function of the *Pistillum* which receives this *Dust*, that it does the Office of the *Parts* of *Generation* in the Female *Animals*; and that what we have before observed, as far as the *Intromission* of the *Dust* into the *Pistillum*, agrees well enough with the *Conception* of the *Fœtus*.

The Nourishment and Growth of the *Embryo* Seed after its *Germ* is made *fœcund*, is agreeable to the Growth of the *Embryo Animal*; the Fruit which encloses it, whether it be *Membraneous, Ligneous*, &c. or whether it be in the Form of a *Capsule, Cod*, or *Siliqua*, or is divided into *few*, or *many*, Cells or Lodgments, that Fruit (as *Malpighius* observes) serves as a *Matrix* to the *Seed*.

From this Description of the *Parts* of a *Flower*, and the Observations upon their Uses, we may draw *two* Consequences.

The *first* is, That those *Parts* in *Plants* which may be termed *Flowers*, are those which perform the *Office* of *Generation*.

Secondly, That after the same Manner as in the *Animal* World, we distinguish between the *Males, Females*, and the *Androgynous*, we likewise discover those *Distinctions* of *Sexes* in *Plants*, which Dr. *Grew* has already touched upon.

We may then conclude that a *Plant* may be termed *Male*, when its *Stamina* do not encompass any *Pistil* or *Stile*; or that the *Stile*, if it has any, is barren, or does not inclose an *Embryo*-Seed: Of this kind, are those Strings or Bunches of *Flowers* which we call *Catkins*, or *Julii*, and the false Blossoms of *Hops*, *Hemp*, *Mercury*, and some others.

On the contrary, the *Female* is easily known by its *Pistils* or *Stiles*, which are not incompassed by *Stamina*, but only guarded with *Petals* or other *Membranes*; and yet are fœcundated by the *Dust* of *Male* Flowers, which either grow upon the same Plant, or upon others of the same Race. This *Fœcundation* is done by the Help of the Wind, which conveys the prolific *Dust* into the *Tubes* of the *Pistils*, when they are advanced to a fit State to receive it; as it is oberable [sic] in the *Walnut*, *Hazle*, *Alder*, *Willow*, *Coniferous* Trees, and *Gourd* kind. *Malpighius* observed these *two* Distinctions in the *Flowers* of the last mentioned *Tribe*, as we may remark by the Figures he has given of them in his *Anatomy* of *Plants*.

The certain Mark by which we may discover the *Androgynous* Flowers, is the ranging of the *Stamina* about the *Pistillum*, whose *Base* or *Body* becomes a Fruit; since we have already remarked, that the *Stamina*, which are the *Male* Parts, will fœcundate the *Pistils* in the same *Flower*, which Part we have observed is found only in the *Female*. There is this only *Difference* between the *Plants* and *Animals* that are *Androgynous*; *Plants* accomplish their *Generation* in themselves without the Help of another Individual of the same *Tribe*; and the Animals, altho' they are endowed with *Organs* agreeable to both Sexes, are yet obliged to seek for one of their own Race to couple with. *Plants* for the most Part bring *Flowers* of this last Species, (that is to say) such *Flowers* as end in *Fruit*.

This Discovery is the Result of those Observations which have been made in the *Anatomy* of *Flowers* and *Fruit*, since it has been judged necessary, that those Parts of *Plants* were the most proper to establish their Characters; and it is not to be doubted but that Time and Industry may disclose to us the *Organs* of the same Uses in those *Plants*, which have been stiled hitherto *Imperfect*, and which will no longer bear that Character when their *Sex* shall be determined.

We cannot in common Justice refuse to give the Honour due to *John Baptist Porta*, for having first observed the *Seeds* in certain *Plants*, which, till his Time, were esteemed *barren*, as in the *Truffle* and *Mushroom*, which has since been confirmed by the curious Remarks of Monsieur *Geoffroy*,[4] Junior; those of Monsieur *Marchand* on the *Agaricus digitatus niger*,[5] and those of Monsieur *le Comte de Marsigli*[6] upon the *Lytophyton*, in whose *Bark* he has found the *Seed*; and we begin likewise to discover them in many *Marine Plants*, but chiefly in the *Fucus*.

Monsieur *Billerer*,[7] Professor of Physic at *Bezançon*, informs me that he has even discovered *Seed* in a River *Sponge*, called *Spongia ramosa fluviatilis*. There is room to believe, that if we were to take a little Pains to examine the *Marine Plants* at different Seasons, we might discover their *Flowers*. or such Parts as acted for them, since Monsieur *de Réaumur* and Signior *Michaeli*, Botanists to the Duke of *Florence*, have already discovered certain Parts which might reasonably be esteemed Dependants of *Flowers*.

This *Distinction* of *Sexes* being established in *Plants*, is one of the most considerable Marks of the *Analogy* between *Plants* and *Animals*; but as it is not only by the *Difference* of *Sexes*, nor by the Use of the *Organs* of *Generation*, that we precisely characterize the different *Animals*; to neither must we be persuaded that these Differences in *Plants* can contribute to distinguish their several Tribes; for in many *Plants*, those Parts which mark out the *Sex* are not easily discovered; and in others the *Flowers* are of so short Duration, that we are not always happy enough to find them in a right Condition for Observation.

APPENDIX

Notes

Introduction

1
J. von Sachs, *Histoire de la botanique du XVIe siècle à 1860* (Paris, 1892), p. 273, and, more recently, P. C. Ritterbush, *Overtures to Biology: The Speculations of Eighteenth-Century Naturalists* (New Haven, Conn., 1964), p. 81.

2
von Sachs, *Histoire*, p. 373.

3
J. Rostand, *Les Origines de la biologie expérimentale et l'Abbé Spallanzani* (Paris, 1951), p. 398.

4
von Sachs, *Histoire*, p. 398.

5
Rostand, *Les Origines*, p. 215.

6
Ritterbush, *Overtures*, pp. 148–52.

7
This is the attitude of most historians of botany. For example, see Raoul Combes, *Historie de la biologie végétale en France* (Paris, 1933), pp. 9–10; and Davy de Virville, *Histoire de la botanique en France* (Paris, 1954), p. 95.

8
M. Foucault, *Les Mots et les choses* (Paris, 1966), p. 139; trans. into English as *The Order of Things* (New York, 1971), p. 127.

9
E. Littré and C. Robin, the article "Végétalité" in the *Dictionnaire de Médecine*, 10th ed. (Paris, 1855), p. 1327a. The term and the definition are borrowed from Auguste Comte; see his *Système de politique positive* (Paris, 1851–1854), vol. 1, p. 594, and vol. 4, p. 219.

10
Noël de Necker, *Phytozoologie philosophique* (Neuwied, 1790), p. 96 (note 59).

Chapter 1

1
Charles Sénebier, Preface to the *Encyclopédie méthodique* (Paris, 1799), p. 1.

2
Ibid., p. 3.

3
Ibid., the article entitled "Physiologie végétale," pp. 221–22.

4
Jean-Jacques Rousseau, "Les Rêveries du promeneur solitaire" (Septième Promenade) in *Oeuvres*, 1793, vol. 26, p. 308.

5
Duhamel du Monceau, Preface to *De l'exploitation des bois* (Paris, 1764), p. 7. This point has been emphasized by F. Dagonet: "The tree is no longer ornamental but rather useful. It is no longer a living thing but a fiber, which it is important to harden, to modify, and even to stretch to the maximum possible length." *Des révolutions vertes* (Paris, 1973), p. 24.

6
François Jacob, *La Logique du vivant* (Paris, 1970), pp. 41–42.

7
Nehemiah Grew, *The Anatomy of Plants (London, 1682)*, preface, unpaginated.

8
Bernard Lamy, *Entretiens sur les sciences*, critical edition, 1966, p. 261.

9
Girolamo Cardano, *Les Livres intitulés de la subtilité* (French trans. of *De subtilitate rerum*, Paris, 1584), p. 196b.

10
Guy de la Brosse, *De la nature, vertu et utilité des plantes* (Paris, 1680), p. 47.

11
Francis Bacon, *Histoire naturelle de M. Francis Bacon* (Paris, 1631), p. 227.

12
Grew, *Anatomy of Plants*, preface, unpaginated.

13
Ibid.

14
Ibid. It was Swammerdam who compared the growth of grass, the caterpillar, and the carnation. This led him to the conclusion that "the same transformations are found in animals and in vegetables, and, though so disparate, these productions of nature are subject to quite simple and uniform general laws." *Histoire naturelle des insectes* (Dijon, 1758), p. 12.

15
Dodart, "Second mémoire sur la fécondité des plantes. Conjectures sur ce sujet," *Mémoires de l'Académie des Sciences*, Paris (1701):247.

16
Swammerdam, *Histoire naturelle des insectes*, p. 24.

17
John Ray, *L'Existence et la sagesse de Dieu manifestées dans les oeuvres de la création* (Utrecht, 1714), pp. 352–53.

18
Sénebier, the article "Physiologie" in *L'Encylopédie méthodique*, p. 221.

19
Grew, *Cosmologia Sacra, or a Discourse of the Universe as It Is the Creator and Kingdom of Good* (London, 1701), p. 18.

20
Aristotle, *Petits Traités d'histoire naturelle (Parva naturalium)*, book II, 468a.

21
Aristotle, *De partibus animalium*, book I, 650a. Hippocrates had already said that "what the earth is to trees, the stomach is to animals; it nourishes, heats, cools; empty, it cools; full, it heats; a smoky earth is warm in winter; the same for the belly." *Oeuvres complètes d'Hippocrate*, new translation by E. Littré, 1962, vol. 5, p. 49.

22
Aristotle, *On the Generation of Animals*, book I, 715b.

23
Cesalpinus, *De plantis*, libris XVI, 1683, p. 2.

24
Duret, *Histoire admirable des plantes* (Paris, 1605), pp. 338–39.

25
Porta, *La Magie naturelle*, French trans. (Rouen, 1680), p. 59.

26
Dodoens, *Histoire des plantes*, French trans. (Antwerp, 1557).

27
John Gerard, *The Herball, or General History of Plants* (London, 1633), p. 647.

28
La Mettrie, "Traité de l'Ame," *Oeuvres philosophiques* (London, 1751), pp. 104–5. See François Bayle, *Dissertationes physicae in quibus principia proprietarum in mistis, Œconomia corporum in plantis et animalium* (The Hague, 1678), p. 54.

29
Grew, *Cosmologia Sacra*, book II, p. 36.

30
Georges Canguilhem, "Modèles et analogies dans la découverte en biologie," in *Etudes d'histoire et de philosophie des sciences* (Paris, 1968), p. 306.

31
Bernard de Jussieu, "De la nécessité d'établir dans la méthode nouvelle des plantes, une classe particulière pour les fougères, à laquelle doivent se rapporter, non seulement les champignons, les agarics, mais aussi les lichens," *Mémoires de l'Académie des Sciences*, Paris (1728): 379.

32
Sénebier, *Encyclopédie méthodique* (Paris, 1799), p. 89.

33
Sénebier, *L'Art d'observer* (Geneva, 1775), vol. 2, p. 93.

34
Ibid., p. 64.

35
Ibid., vol. 1, p. 88.

36
Ibid., vol. 2, p. 86.

37
Ibid., vol. 1, p. 55.

38
Ibid., p. 10.

39
J. von Sachs, *Histoire de la botanique*, p. 254.

Chapter 2

1
An expression I have borrowed from the author of the article "Végétation" in the *Dictionnaire de Trevoux*, 1771, vol. 8, p. 308a: "The system of vegetation is the mechanism by which it operates." Jaucourt was the author of the article "Plante" in the

Encyclopédie, 1781, vol. 26, pp. 102–3: "It seems that the mechanism of *plants* is quite similar to the mechanism of animals: the parts of *plants* seem to bear a fixed analogy to the parts of animate bodies; and the vegetable economy seems to be modeled after the animal economy."

2
J. D. Major, *Dissertatio botanica, de planta monstrosa Gottorpiensi* (Schelwig, 1665), II, unpaginated.

3
Ibid., III.

4
Mariotte, "De la végétation des plantes," *Œuvres* (The Hague, 1740), vol. 1, pp. 130–31.

5
Perrault, "De la circulation de la sève dans les plantes," *Essais de physique* (Paris, 1680–81), vol. 1, p. 180.

6
Dedu, *De l'âme des plantes, de leur naissance, de leur nourriture et de leur progrès, Essai de physique* (Paris, 1682), pp. 32–33.

7
Mariotte, "De la végétation des plantes," pp. 129–30. Leeuwenhoek made the same assumption; see "An abstract of a letter from Mr. Leeuwenhoek, concerning the appearance of several woods, and their vessels," *Philosophical Transactions* 13(1683):121. So did La Hire; see "Sur la cause de l'élévation du suc nourricier dans les plantes," *Histoire de l'Académie Royale*, Paris 2(1686):185–86, and "Expériences servant d'éclaircissement à l'élévation du suc nourricier dans les plantes," *Mémoires de l'Académie des Sciences*, Paris X(1666–1699):317–19.

8
Perrault, "De la circulation de la sève," vol. I, p. 196.

9
Ibid., p. 177.

10
Major, *De planta monstrosa Gottorpiensi*, part 33, sec. II.

11
Perrault, "De la circulation de la sève," vol. I, p. 240.

12
Grew, *Anatomy of Plants* (London, 1682), book II, sec. 23, p. 84; same remark, book III, sec. 21, p. 131. Grew had adopted this theory, but without developing all of its implications, as early as 1672 in *Anatomy of Vegetables, Begun*, chap. II, p. 52: "So that as in an *Animal Body* there is no instauration or growth of Parts made by the Blood only, but the *Nervous Liquor* is also thereunto

assistant; so it is here: the *sap* prepared in the *Cortical Body*, is as the Arterious; and that part thereof prepared by the *Lignous*, is as the *Nervous Liquor*."

13
Grew, *Anatomy of Plants*, book II, p. 86.

14
Malpighi, "Anatomes plantarum idea," *Opera omnia* (London, 1686), p. 14.

15
Ibid., p. 3. Cuvier attributes the discovery of the tracheae to Henshaw: "In 1661, Henshaw, too, used magnifying glasses and made the beautiful discovery of the tracheae of vegetables." *Histoire des sciences naturelles* (Paris, 1843), vol. IV, p. 56. Undoubtedly Henshaw did observe this part, but he did not give it the name trachea (owing to the similarity with the tracheal artery of animals). What is more, he says nothing about their use. On this point, see Thomas Birch, *The History of the Royal Society of London* (London, 1756–57), vol. I, July 31, 1661, p. 37: "Mr. Henshaw, Thomas, exhibited the spirales of nut-trees, shewing, that they grow snail-wise."

16
Ibid., p. 5.

17
Grew, *Anatomy of Plants*, vol. III, sec. 29, p. 111.

18
Mariotte, "De la végétation des plantes," vol. I, pp. 129–30.

19
Malpighi, "Anatomes plantarum idea," p. 38.

20
E. F. Geoffroy, *Thèse soutenue aux ecoles* (Paris, 1705), p. 16. Similarly, for Duhamel du Monceau, "the air, by means of its rarefaction and condensation, must be regarded as the prime mover of the sap; similarly, the heart, by its motion of compression and expansion, is the first principle of the circulation of the blood." "Recherches physiques de la cause du prompt accroissements des plantes dans les temps de pluies," *Mémoires de l'Académie des Sciences*, Paris (1729):356.

21
Duhamel du Monceau, "Recherches" (see preceding note), p. 357.

22
Bazin, *Observations sur les plantes et leur analogie avec les insectes* (Strasbourg, 1741), p. 96.

23
Ibid., p. 100.

24
Jethro Tull, *The Horse Hoeing Husbandry: or an Essay on the Principles of Tillage and Vegetation* (London, 1733), chap. 1, p. 7.

25
Ibid., p. 8.

26
Anonymous, "Extrait d'un discours sur la respiration des plantes," *Journal de Trevoux* (1712):908.

27
Ibid., pp. 902–3.

28
Bonnet, *Recherches sur l'usage des feuilles dans les plantes* (Göttingen and Leyden, 1754), mem. I, p. 52.

29
Ibid., pp. 28–9.

30
Bertholon, *De l'électricité des végétaux* (Paris, 1783), p. 217.

31
Gesner, *Dissertationes physicae de vegetabilibus quarum prior partium vegetationis structuram* (Leyden) thesis XVII, p. 75. See also La Metherie, *Vues physiologiques sur l'organisation animale et végétale* (Amsterdam and Paris, 1780), p. 30.

32
Bonnet, *Recherches sur l'usage des feuilles*, p. 62.

33
H. B. de Saussure, *Observations sur l'écorce des feuilles et des pétales* (Geneva, 1762), p. 2. See also Duhamel du Monceau, "Anatomie de la poire," *Mémoires de l'Académie des Sciences*, Paris (1730):299–324.

34
Renéaume, "Observations sur le suc nourricier des plantes," *Mémoires de l'Académie des Sciences*, Paris (1707):276–77.

35
Musschenbroek, *Cours de physique expérimentale et mathématique*, French trans. (Leyden, 1769), vol. 3, p. 297. See also Gersten, *Tentamina systematis novi ad mutationes barometri ex natura elateris aeri demonstrandas, cui adjecta sub finem dissertatio Roris decidui errorem antiqum et vulgarem per observationes et experimenta nova excutiens* (Frankfurt, 1733).

36
Guettard, "Mémoires sur les corps glanduleux des plantes, leurs filets ou poils, et les matières qui suintent des unes et des autres," *Mémoires de l'Académie des Sciences*, Paris (1745):267. See Ob-

servations sur les plantes (Paris, 1747), vol. 1, preface, p. 36, and Hertel, *Dissertationem physiquam de plantarum transpirationem* (Leipzig, 1735).

37
J. Hill, *The Vegetable System* 2nd ed. (London, 1770), vol. 1, chap. VII, p. 32.

38
Duhamel du Monceau, "Sur le développement et la crue des os des animaux," *Mémoires de l'Académie des Sciences*, Paris (1742):369. On this question, see L. Plantefol, "Duhamel du Monceau," *Dix-huitième siècle* 1(1969):123–37.

39
Stephen Hales, *Vegetable Staticks*, first published in 1727; citations are to 1961 London reprint: pp. xxi-ii.

40
Ibid., p. xxxii.

41
On this point, see Hales, *Haemastatique, ou la Statique des animaux* (Geneva, 1744), introduction, pp. xvi-xvii.

42
Hales, *Vegetable Staticks*, p. 6. In his computations, Hales made use of the work of J. Keill, author of the *Medicina statica Britannica* (London, 1718).

43
Ibid.

44
Ibid.

45
Ibid.

46
J. Woodward, "Some thoughts and experiments concerning vegetation," *Philosophical Transactions* 21(1699):213.

47
Hales, *Vegetable Staticks*, chap. III, p. 59, Experiment XXV.

48
Ibid., Experiment XXV.

49
Guettard, "Second mémoire sur la transpiration insensible des plantes," *Mémoires de l'Académie des Sciences*, Paris (1749):272.

50
Guettard, "Mémoire sur la transpiration insensible des plantes," *Mémoires de l'Académie des Sciences*, Paris (1748):570.

51
Ibid., p. 577.

52
Guettard, "Second mémoire," p. 275.

53
Gouan, *Discours sur la cause du mouvement de la sève dans les plantes* (Montpellier, 1803), p. 26.

54
Guettard, "Second mémoire," p. 269.

55
Renéaume, "Observations sur le suc nourricier des plantes," *Mémoires de l'Académie des Sciences*, Paris (1707):283.

56
Parent, "Sur la nourriture des plantes," *Histoire de l'Académie Royale des Sciences* (1711):43.

57
Sarrabat, *Dissertation sur la circulation de la sève* (Bordeaux, 1732), pp. 20–21. It is true that this procedure was used earlier by Magnol, but only to prove that Perrault was wrong in arguing that the sap circulates: "He soaked a tuberous stem flower overnight in juice of *solanum raceofum* mixed with a little water. This juice is the color of gum and the stem took on a beautiful pink color. Apparently the juices that caused this change of color and therefore fed the plant most intimately cannot have been changed or affected very much." "Sur la circulation de la sève dans les plantes," *Histoire de l'Académie Royale des Sciences*, Paris (1709):48.

58
Bonnet, *Recherches sur l'usage des feuilles*, mem. V, p. 248. Reichel also soaked a number of plants in wood decoctions and observed that the liquor rose not only through the ligneous fibers but also through the tracheae. See *De vasis plantarum spiralibus* (Leipzig, 1756), pp. 24–25, Experiment I.

59
Duhamel du Monceau, "Sur le développement et la crue des os," p. 360. See Hales, *Vegetable Staticks*, chap. VII, Experiment CXXIII.

60
Hales, *Vegetable Staticks*, chap. VII, p. 193.

61
The cortex is joined to the wood "as the skin of the body is to the flesh." Grew, *Anatomy of Plants*, book III, chap. III, sec. 2, p. 129.

62
Duhamel du Monceau, "Sur la réunion des plaies des arbres et des

animaux; et sur les greffes ou incisions tant végétales qu'animales," *Mémoires de l'Académie des Sciences*, Paris (1746):332.

63
Hales, *Vegetable Staticks*, chap. IV, p. 83, Experiment XLVI.

64
Mustel, *Traité théorique et pratique de la végétation* (Paris and Rouen, 1781–84), vol. II, book III, chap. XIII, pp. 157–8. See "New observations upon vegetation by Mr. Mustel," *Philosophical Transactions* 63(1773):126–36. In the spring, if a notch is cut in a tree, the lower part of the cortex is the first to be moistened by sap. "These facts are also important points against the doctrine of circulation." Walker, "Experiments on the motion of the sap in trees," *Transactions of the Royal Society*, Edinburgh, I, sec. 7 (1788):39.

65
Mayow, *De respiratione*, 1668, p. 43; English translation by T. S. Patterson, "John Mayow in contemporary setting. A contribution to the history of respiration and combustion," *Isis* 15(1931):47–96. Similarly for Hooke: "Mr. Hooke reported that he had sown some lettice-seed upon earth in the open air, and at the same time upon other earth in a glass-receiver ... which was afterwards exhausted of air; that the seed exposed to the air was grown up an inch and a half within eight days; but that in the exhausted receiver not at all: both which were produced and shewn the society." Birch, *The History of the Royal Society of London, 1756–1757*, vol. II, p. 54.

66
Hales, *Vegetable Staticks*, chap. V, p. 86. See Nieuwentyt, *L'Existence de Dieu demontrée par les merveilles de la nature* (Amsterdam and Leipzig, 1760), book II, chap. VIII, p. 371. Christian Wolf repeated Nieuwentyt's experiments and also showed with the aid of a pneumatic machine that the tracheae contain air and that they are true respiratory organs. *Vernünftige Gedanken von den Wirkungen der Natur* (Halle, 1723).

67
Duhamel du Monceau, "Recherches sur la formation des couches ligneuses dans les arbres," *Mémoires de l'Académie des Sciences*, Paris (1751):31–32.

68
Ibid., p. 33.

69
Ibid., p. 35.

70
J. B. Bucquet, *Introduction à l'étude des corps naturels, tirés du règne végétal* (Paris, 1773), vol. 2, p. 251.

71
Duhamel du Monceau, *Eléments d'agriculture* (Paris, 1762), vol. 1, book I, chap. VI, p. 44.

72
Sarrabat, *Dissertation sur la circulation de la sève*, pp. 58–59.

73
Fouquet, the article "Sécrétion" in the *Encylopédie*, vol. XXX, p. 213.

74
Bradley, "Observations and experiments relating to the motion of the sap in vegetables," *Philosophical Transactions* 29(1716):488. Parsons makes a similar remark; see *Philosophical Observations on the Analogy between the Propagation of Animals and Vegetables* (London, 1752), note X, pp. 140–1.

75
Mariotte, "De la végétation des plantes," vol. I, p. 130.

76
Ibid., pp. 140–41.

77
Tull, *Horse Hoeing Husbandry*, p. 103.

78
Bradley, *Le Calendrier des jardiniers* (Paris, 1743), p. 69.

79
Fouquet, the article "Sécrétion," p. 490.

80
Parsons, *Philosophical Observations*, p. 142.

81
Mariotte, "De la végétation des plantes," part 3, p. 145.

82
"All the flavors," says Aristotle, "that may be found in the pericarps may also be found in the earth." *Minor Treatises on Natural History* 4, 441a, 30–31.

83
Tull, *Horse Hoeing Husbandry*, chap. VI, p. 104.

84
Van Helmont, *La Source de la médecine* (Lyon, 1670), chap. XV, p. 101. He used a willow that weighed five pounds at the beginning of the experiment and 169 pounds after five years of vegetation in 200 pounds of earth. The earth had lost only two ounces of its weight, however. Boyle planted a squash seed in a vase containing a known quantity of soil that had been dried in an oven; he watered this earth and obtained a plant that weighed more than fourteen pounds, but the earth had lost almost no weight. The water was therefore the unique cause of the growth of plants.

"The Sceptical chymist," in *The Works of the Honourable Robert Boyle* (London, 1744), vol. I, part 2, pp. 312–13.

85
Parsons, *Philosophical Observations*, chap. IV, p. 141.

86
Bradley, *A Philosophical Treatise of Agriculture*, by G. A. Agricola. The whole revised and compared with the original (London, 1721), chap. IV, p. 36 [passage cited here translated from French—trans.].

87
Tull, *Horse Hoeing Husbandry*, chap. XVI, p. 109.

88
Mustel, *Traité théorique et pratique de la végétation* vol. II, book IV, chap. IX, p. 257.

89
Buffon, *Histoire naturelle générale et particulière* (Paris, 1749), chap. I, p. 8.

90
Hales, *Vegetable Staticks*, chap. IV, p. 77. Similarly, for Antoine-Laurent de Jussieu, "the animal moves spontaneously from one place to another, seeks food over some distance, carries the food to its mouth, chews it, digests it; the plant, on the other hand, is attached to the soil and immobile and feeds solely on the juice of the nearest earth. The lack of movement and circulation are compensated, however, by a constant suction and by the receipt of a larger quantity of juice." *An Œconomiam Animalem inter & Vegetalem Analogia?* (Paris, 1770), p. 11.

91
Descartes, "Les principes de la philosophie," *Œuvres et lettres* (Paris, 1953), pp. 583–84.

92
F. Jacob, *La Logique du vivant*, p. 45 (English trans., p. 36).

93
Huygens, "Le Cosmotheros," *Œuvres completes de Christian Huygens* (The Hague, 1944), vol. XXI, p. 700.

94
Nieuwentyt, *L'Existence de Dieu*, book II, chap. VIII, pp. 355–56.

95
J. Ray, *L'Existence et la sagesse de Dieu manifestées dans les oeuvres de la creation*, French trans. (Utrecht, 1714), p. 114.

96
Montesquieu, "Observations sur l'histoire naturelle," *Œuvres de Montesquieu* (Paris, 1819), vol. VII, pp. 184–85.

97
Mariotte, "De la végétation des plantes," p. 126.
98
Montesquieu, "Observations," vol. VII, p. 189.
99
Boerhaave, Eléments de chimie (The Hague, 1748), vol. I, part II, p. 75. F. Courtès has emphasized this point: "Boerhaave proposed a unique figure in relation to which plants and animals appeared as opposites: to get from one to the other required a reversal, as one might do with the finger of a glove." In E. Kant, Rêves d'un visionnaire (Paris, 1967), p. 139, n. 7.
100
Concerning Buffon as the translator of Hales, see L. Hanks, *Buffon avant l'"histoire naturelle"* (Paris, 1966), part 2, chap. I, pp. 73–88.
101
Buffon, Histoire naturelle, générale et particulière (Paris, 1749), vol. IV, p. 9.
102
Ibid., vol. I, p. 443. Flourens, who stressed this point, insisted on Buffon's Cartesianism; see Histoire des travaux et des idées de Buffon (Paris, 1870), pp. 115–23.
103
La Mettrie, "L'Homme-plante," Œuvres philosophiques (London, 1751), pp. 254–55.
104
Ibid., p. 262.
105
Ibid., p. 254.
106
Diderot, the article "Animal" in the Encylopédie, p. 672.
107
Ibid., p. 676.
108
Ibid., p. 175.
109
Foucault, Les Mots et les choses, p. 173.
110
Boerhaave, Eléments de chimie, vol. I, part II, p. 75.
111
Ibid., p. 74.
112
Ibid., p. 66.

113
Vicq d'Azir, "Premier Discours sur l'anatomie," *Œuvres* (Paris, 1805), vol. IV, p. 18. The following oppositions also followed as a result: organic/inorganic, growth by intussusception/growth by juxtaposition. On this point, see Daudin, *De Linné à Jussieu, Méthode de la classification et l'idée de série* (Paris), p. 177ff.

114
Hoffmann, *Médecine raisonné* (Paris, 1739), vol. I, book I, chap. II, p. 115.

115
Kant, *Rêves d'un visionnaire*, p. 63.

116
Hartsoeker, *Eclaircissements sur les conjectures physiques* (Amsterdam, 1710), vol. II, p. 115.

117
La Quintinie, *Instructions pour les jardins fruitiers et potagers avec un traité des orangers, et des réflexions sur l'agriculture*, new ed., revised, corrected, and augmented (Paris, 1756), vol. II, chap. VI, p. 296.

118
Ibid., chap. VIII, p. 302.

119
Ibid., pp. 301–2.

Chapter 3

1
I am borrowing this expression from Michel Foucault, who has observed that prior to the nineteenth century "sexuality was considered to be a sort of supplementary apparatus thanks to which the individual, once a certain stage of growth was achieved, shifted to a new mode of growth, by multiplication rather than increase in size. Sexuality was a sort of dynamo of growth." *Thales* 13(1969):91, special issue on Georges Cuvier.

2
In the late eighteenth and early nineteenth centuries, some botanists denied the existence of sexuality in lower plants (mosses and ferns); others argued for the opposite view: "The supporters of the latter hypothesis are known as sexualists, in contrast to the others, who are known as agamists." Antoine-Laurent de Jussieu et de Mirbel, "Rapport sur le Mémoire de M. Desvaux sur les lycopodiacées, et monographie de cette famille," *Journal de physique* 76(1813):323. The term *agamists* could also apply to botanists who denied the existence of sexuality in the higher plants.

3
J. Ray, *Stirpium Europaerum extra Britannias nascentium Silloge* (London, 1694), preface, unpaginated. Camerarius made the same remark; see "Semina mori subventanea," *Ephemerid. Germ.*, 1691, p. 213. He observes that a female mulberry bears fruit even when there is no male mulberry nearby. He also notes that bushes that grow in these conditions contain only hollow and empty seeds, which he compares to the sterile eggs sometimes laid by birds. The infertile eggs produced without the influence of the male were generally called "eggs of the wind," because birds seemed to receive a fertilizing breeze from the west wind in the springtime. Concerning this myth, see C. Zirkle, "Animals impregnated by the wind," *Isis* 25(1936):95–130.

4
Camerarius, *De sexu plantarum epistola* (Tübingen, 1694), p. 20, cited in J. von Sachs, *Histoire de la botanique*, p. 401.

5
Ibid., p. 24.

6
Ibid., pp. 38–39. Camerarius makes the same observation about another species: "Garden mercury also produced seeds, but none of them germinated when the plant was isolated from male stalks." Ibid., p. 39.

7
Bradley, *New Improvements of Planting and Gardening both philosophical and practical, explaining the motion of the sap and the generation of plants* (London, 1717–18), vol. I, p. 21. Camerarius had already done similar experiments with ricinus and maize. See *De sexu plantarum epistola*, p. 38, cited in J. von Sachs, *Histoire de la botanique*, book III, chap. I, pp. 400–1.

8
Bradley, *New Improvements*, p. 20.

9
G. Bonnier, *Le Monde végétal* (Paris, 1913), p. 9–10.

10
Miller, article "Generation," *The Gardeners and Florists Dictionary, or a Complete System of Horticulture* (London, 1724), unpaginated.

11
Blair, *Botanick Essays* (London, 1720), Essay IV, p. 247.

12
Bradley, *New Improvements*, vol. I, p. 12.

13
Vaillant, *Discours sur la structure des fleurs, leurs differences et l'usage de leurs parties* (Leyden, 1718), p. 6.

14
Ibid., p. 14.

15
Miller, "Letter to Mr. Bradley, October 6, 1721," in *A General Treatise of Husbandry and Gardening* (London, 1724), vol. II, p. 16.

16
Bradley, *New Improvements*, p. 12.

17
Bradley, *A Philosophical Account of the Works of Nature*, 1721, p. 113.

18
C. Geoffroy, "Observations sur la structure et l'usage des principales parties des fleurs," *Mémoires de l'Académie des Sciences*, Paris (1711): p. 211. The same for Linnaeus: "The parent is a hermaphrodite, as commonly occurs in plants." *Systema Naturae* (Leyden, 1735), part III, p. 18.

19
Bradley, *New Improvements*, p. 15.

20
P. Dudley, "Observations on some plants in New England with remarkable instances of the nature and power of vegetation," *Philosophical Transactions*, no. 385 (1724):199.

21
C. Geoffroy, "Observations sur la structure et l'usage des principales parties des fleurs," *Mémoires de l'Académie des Sciences*, Paris (1711):229. The same for S. Morland: "This flour is a stockpile of seminal plants, one of which must be introduced into each egg to make it prolific." "Some new observations upon the parts and uses of flower in plants," *Philosophical Transactions* no. 287 (1703):1475.

22
Vaillant, *Discours sur la structure des fleurs*, p. 16.

23
Leibniz, *New Essays Concerning Human Understanding*, trans. A. G. Langley (Chicago, 1916), book III, chap. VI, sec. 23, p. 347. In a letter to Leibniz, Johann Heinrich Burckhard adopted the system of plant sexes: see *Epistola ad Illustrem . . . dominum Godofredum Guililmum Leibnitium, . . . qua characterem plantarum naturalem nec a radicibus, nec ab aliis plantarum partibus . . .* (Helmstedt, 1701).

24
Linnaeus, "Dissertation sur les sexes des plantes," *Journal de physique*, Rozier 32(1788):453.

25
Logan, *Experiments and Considerations of Generation of Plants* (London, 1747), p. 11.

26
Gleditsch, "Essai d'une fécondation artificielle," *Mémoires de l'Académie des Sciences*, Berlin (1749):107. See also "Relation de la fécondation artificielle d'un palmier femelle, réitérée pour la troisième fois, et avec un plein succès," *Mémoires de l'Académie des Sciences*, Berlin (1767):3–19. In this period travelers were still describing this practice. On this point, see Hasselquist, "Lettre d'Alexandrie du 18 mai 1750," in *Voyage dans le Levant*, French trans., 1769, p. 50; and C. G. Ludwig, *Dissertatio de sexu plantarum* (Frankfurt, 1782), p. 25.

27
Logan, *Experiments and Considerations of Generation of Plants*, p. 11. A variant: Linnaeus rubbed one of the three stigmata of *Clutia tenella* and covered the other two so that the fertilizing dust could not touch them: "The fruit achieved its ordinary plumpness, and when I cut it transversally I found a plump seed in one locule, while the two others were empty." "Dissertation sur les sexes des plantes," p. 454.

28
Gleichen, *Découvertes les plus nouvelles dans le règne végétal* (Nuremberg, 1776), p. 34.

29
Linnaeus, "Economie de la nature," in *L'Equilibre de la nature*, 1972, p. 69.

30
Adanson, *Famille des plantes* (Paris, 1763), vol. 1, p. 123.

31
Adanson, "Examen de la question si les espèces changent parmi les plantes: nouvelles expériences tentées à ce sujet," *Mémoires de l'Académie des Sciences*, Paris (1769):48.

32
Gleditsch, "Dissertation sur un pommier à tige basse," *Mémoires de l'Académie des Sciences*, Berlin (1754):86–7.

33
Gleditsch, "Relation de la fécondation artificielle," *Mémoires de l'Académie des Sciences*, Berlin (1767):5. See also Gleditsch, "Exposition d'une fécondation artificielle des huîtres et des saumons, qui est appuyée sur des expériences certaines, faites par un habile naturaliste," ibid., 1764, p. 52. Concerning collaboration of the

two kingdoms (animals accomplishing the transport of the seed), see *L'Equilibre de la nature*, pp. 71–72; and the introduction by C. Limoges, pp. 7–22.

34
Gleditsch, "Dissertation sur un pommier à tige basse," p. 88.

35
Bonnet, "Idées sur la fécondation des plantes," *Œuvres*, 1781, vol. X, p. 86.

36
Gleichen, *Découvertes les plus nouvelles dans le règne végétal*, 1770, LXXXI, p. 43.

37
Fougeroux de Bondary, "Dissertation sur la fécondation des plantes," *Journal de physique*, Rozier 19(1775):26–27.

38
Needham, *Nouvelles découvertes faites avec le microscope*, trans. from the English (Leyden, 1747), chap. VII, p. 84.

39
Ibid., p. 91.

40
Jacob, *La Logique du vivant*, p. 74 (English trans. p. 62).

41
Bulliard, *Histoire des champignons de la France* (Paris, 1791–1812), p. 7 ff; see also Palisot de Beauvois, *Prodrome des 5ᵉ et 6ᵉ familles de l'Aetheogamie* (Paris, 1805), p. 114: "Plants such as mosses and ferns which have leaves must have an embryo, although the smallness of their seed does not allow us to make them [sic] out."

42
Bernard de Jussieu, "Histoire d'une plante connue par les botanistes sous le nom de Pilularia," *Mémoires de l'Académie des Sciences*, Paris (1740):243.

43
Réaumur, "Description des fleurs et des graines de divers fucus, et quelques autres observations physiques sur ces mêmes plantes," *Mémoires de l'Académie des Sciences*, Paris (1711):282–302.

44
Hedwig, *Theoria generationis et Fructificationis Plantarum cryptogamicarum Linnaei* (Petropoli, 1784), chap. IV, p. 82. See also Dillenius, *Historia muscorum* (Oxford, 1741).

45
Bulliard, *Histoire des champignons*, pp. 36–37.

46
Bernard de Jussieu, "Histoire d'une plante," *Mémoires de l'Académie des Sciences*, Paris (1740):255.

47
Guettard, "Observations par lesquelles on détermine le caractère générique de la plante appelée Marsilea, plus exactement qu'il ne l'a été jusqu'à présent," *Mémoires de l'Académie des Sciences*, Paris (1742):548.

48
Correa de Serra, "On the fructification of the submersed Algae," *Philosophical Transactions* (1796):505. See also J.-F. Lamouroux, *Dissertation sur plusieurs especes de Fucus* (Agen, 1805), p. xxvi: "The mucilaginous matter contained in the tubers is the only agent that helps to fertilize the seeds."

49
Maupertuis, "Vénus physique," *Œuvres de Maupertuis* (Lyon, 1756), vol. II, p. 64. Maupertuis is right to say that both animalculists and ovists make reference to the seed of plants. On the animalculists, see Gardeen of Aberdeen, "A discourse concerning the modern theory of generation," *Philosophical Transactions*, no. 192 (1691):476. On the ovists, see Fontenelle, "Sur la génération de l'homme par les oeufs,"·*Histoire de l'Académie Royale des Sciences* (Paris, 1701), p. 38.

50
Hartsoeker, *Essai de dioptrique* (Paris, 1694), pp. 232–33.

51
Grew, *Anatomy of Plants*, book IV, p. 172; and Malpighi, "Anatomes plantarum idea," *Opera omnia*, p. 56.

52
Tournefort, "Introduction à la botanique," in *Tournefort* (Paris: Muséum national d'histoire naturelle, 1957), p. 302.

53
Ibid., p. 302.

54
Spallanzani, cited by Bonnet, in letter of 27 November 1777 (letter XXXIX), *Œuvres*, vol. XII, p. 270. See also Spallanzani, *Expériences pour servir à l'histoire de la génération des animaux et des plantes* (Pavia, 1787), vol. III, p. 327: "These results do not agree with my observations of animals, in which the fetus can be seen prior to fertilization."

55
Alston, *Dissertation on the Sexes of Plants*, pp. 254–55 [trans. from French].

56
Reynier, "Résultat de quelques expériences relatives à la génération des plantes," *Journal de physique*, Rozier 31(1787):327. In regard to Möller, see "Muthmassliche Gedanken von dem Staube der Pflanzen während der Bluthe," *Hamburgerisches Magazin*, Hamburg and Leipzig II (1748):454–76. Note in passing that the opposite results were obtained by the sexualists. See, in particular, Kastner, "Anmerkungen uber die Muthmasslichen Gedanken von dem Staube der Pflanzen," *Hamburgerisches Magazin* III (1752):11–24. See also A. de Marti, *Experimentos y observationes sobre los sexos y fecundacion de las plantas* (Barcelona, 1791).

57
Spallanzani, "Expériences sur la génération," *Œuvres*, 1787, vol. III, p. 374.

58
Ibid., pp. 397–8.

59
Spallanzani, *Expériences*, chap. V, p. 373.

60
Ibid., p. 401. Möller had earlier made use of the same argument to interpret an experiment done by Bradley: "Since, in my opinion, it is through these anthers [the anthers or summits of the tulip] that the excretion of certain material takes place, and since this material is presumed to be superfluous to the seed, Bradley erred in removing these parts, since this interrupted this very necessary excretion, and for this reason alone, the semen was unable to reach its full perfection." "Fortsetzung der muthmasslichen Gedanken vom Bluhmenstaube," *Hamburgerisches Magazin*, Hamburg and Leipzig III (1752):427.

61
Alston, *Dissertation on the Sexes of Plants*, p. 264.

62
Ibid., p. 265.

63
Ibid., p. 271.

64
Ibid., p. 308.

65
W. Smellie, *The Philosophy of Natural History* (Edinburgh, 1790–99), vol. I, p. 246.

66
A.-P. de Candolle, *Physiologie des plantes* (Paris, 1832), vol. II, p. 513. See A. W. Henschel, *Von der Sexualität des Pflanzen* (Bres-

low, 1820); and F. Schelver, *Kritik der Lehre von den Geschlechtern der Pflanzen* (Heidelberg, 1822).

67
Fougeroux de Bondary, "Mémoire sur la fécondation des plantes," p. 30.

68
Bonnet, Letter XXXIV, 29 Nov. 1777, *Œuvres*, vol. XII, p. 290.

69
Smellie, *Philosophy of Natural History*, vol. I, p. 254.

70
Bonnet, letter XLIV, 24 Feb. 1781. *Œuvres*, vol. XII, p. 399.

71
Ibid., pp. 402–3.

72
The expression is from Joseph de Necker, *Phytozoologie philosophique*, p. 8.

73
J. de Necker, "Eclaircissements sur la propagation des filicées," *Mémoires de l'Académie des Sciences*, Mannheim (1775):275.

74
Joseph de Necker, *Physiologie des corps organisés*, 1775, p. 156.

75
Ibid., pp. 155–56.

76
On this point, see Gaertner, *De fructibus et seminibus plantarum* (Stuttgart, 1788–1807), vol. I, preface.

77
Bachelot de la Pylaie, *Etudes cryptogamiques* (Paris, 1815, p. 7.

78
Alston, *A Dissertation on Botany* (London, 1754), p. 41.

79
Adanson, *Famille des plantes*, vol. I, p. 32.

80
Antoine de Jussieu, *Du rapport des plantes avec les animaux tiré de la différence de leurs sexes*, Manuscript 284 (Muséum d'histoire naturelle).

81
Erasmus Darwin, *Zoonomia, or the laws òf organic life* (London, 1801), vol. I, sec. XIII, p. 175. To this he adds the famous example of the flowers of the vallisneria: "The male flowers of the vallisneria come even closer to apparent animality, for they detach themselves from the plant and float on the surface of the waters to meet their females," p. 142.

82
Linnaeus, *Systema naturae*, clavis systematis sexualis.

83
Bonnet, *L'Amour végétal ou les noces des plantes*, 2nd ed. (Paris, 1809), p. 75.

84
Desfontaines, "Observations sur l'irritabilité des organes sexuels d'un grand nombre de plantes," *Mémoires de l'Académie des Sciences*, Paris (1787):473.

85
Ibid., p. 474.

86
Koelreuter, *Vorläufige Nachricht von einigen das Geschlecht der Pflanzen betreffenden Versuchen und Beobachtungen* (Leipzig, 1791), p. 21, cited in P. Knuth, *Handbook of Flower Pollinization* (Oxford, 1906), vol. I, p. 1.

87
Ibid., p. 23, cited in Knuth, p. 2.

88
Ibid., p. 24, cited in Knuth, p. 2. Koelreuter also gathered nectar from a great many flowers, allowed the liquid to evaporate, and found that the nectar contained a sort of honey with a pleasant taste.

89
Sprengel, *Das entdeckte Geheimniss der Natur im Bau und in der Befruchtung der Blumen* (Berlin, 1793), p. 1, cited in J. von Sachs, *Histoire de la botanique*, pp. 430–31.

90
Koelreuter, *Vorläufige Nachricht*, p. 2, cited in Knuth, p. 4.

91
Ibid., p. 19, cited in Knuth, p. 7. Koelreuter also observed dichogamy in the iris and the œnothera: "Pollination is accomplished entirely by insects, for the anthers open long before the stigma shows itself." *Vorläufige Nachricht*, pp. 34–35, cited in Knuth, p. 3.

92
Ibid., p. 17, cited in Knuth, p. 6.

93
Ibid., p. 19, cited in Knuth, p. 7.

94
Ibid., p. 43, cited in Knuth, p. 8.

95
Knight, "An account of some experiments on the fecundation of vegetables," *Philosophical Transactions* 89(1799):203.

96
Koelreuter, *Vorläufige Nachricht*, p. 29, cited in R. Olby, *Origins of Mendelism* (London, 1966), p. 150.
97
Jacob, *La Logique du vivant*, chap. I, p. 81 (English trans., p. 69).
98
Koelreuter, *Vorläufige Nachricht*, p. 52, cited in Olby, *Origins of Mendelism*, p. 154.
99
Maupertuis, "Vénus physique," vol. II, p. 86.
100
Condorcet, "Eloge de M. Duhamel," in *Eloges des Academiciens* (Paris, 1799), vol. III, p. 194.
101
Rousseau, "Les Rêveries du promeneur solitaire (VIIe promenade)," *Œuvres*, 1793, vol. XXVI, p. 319.
102
Ibid., p. 310.
103
Rousseau, "Lettres élémentaires sur la botanique, 22 août 1771," in ibid., vol. V, p. 8. For Trembley, field work has the same function: "One of life's pleasantest occupations, it offers the soul objects most apt to keep at bay the passions which agitate it and lead it astray; and maintains the calm that enables the soul to admire the beauties of nature and taste the appeal of virtue. The countryside has always been viewed as the abode of innocence." *Instruction d'un père à ses enfants sur la nature et sur la religion* (Geneva, 1775), vol. I, disc. IX, pp. 89–90. On Rousseau and herborization, see Sir Gavin de Beer, "Jean-Jacques Rousseau: Botanist," *Annals of Science* 10(1954):189–223.
104
Rousseau, "Les Rêveries du promeneur solitaire," p. 306.
105
Bonnet, *L'Amour végétal ou les noces des plantes*, Discours préliminaire, p. xiii.
106
Ritterbush, *Overtures to Biology*, p. 119.
107
Linnaeus, *Sponsalia plantarum*, vol. I, p. 104.
108
Smellie, *Philosophy of Natural History*, vol. I, sec. II, p. 248. J. G. Siegesbeck also opposed Linnaeus' sexual classification: "How forced, deceptive, fallacious, and, worse, foolish, are the foundations on which such a method is based." *Botanosophiae verioris*

brevis sciagraphia (Petersburg, 1737), p. 48. And Pontedera attacked the sexualists directly. "He forgets himself," says Gleichen, "in his zeal to attack the defenders of the system of the sexes, to the point of regurgitating insults, calling them chatterers, pseudobotanists, and animals." *Découvertes les plus nouvelles dans le règne végétal* (Nuremberg, 1770), p. 39.

109
Rousseau, "Les Rêveries du promeneur solitaire," p. 308.

110
Linnaeus, *Sponsalia plantarum*, pp. 104–5.

111
Rousseau, "Les Rêveries du promeneur solitaire," pp. 310–11.

112
Gleichen, *Découvertes les plus nouvelles*, p. 40.

113
Rousseau, "Les Rêveries du promeneur solitaire," p. 263.

114
Rousseau, "Les Confessions," vol. XXIII, book I, p. 30.

115
Rotheram, *The Sexes of Plants Vindicated* (Edinburgh, 1790), p. 32. "Pardon my religious sentiments," adds Rotheram for the benefit of his correspondent, William Smellie. [The cited passage is translated from the French—trans.]

116
Pultney, *Historical and Biographical Sketches of the Progress of Botany in England* (London, 1790), vol. I, p. 145. This argument has recently been revived by R. C. Olby, *Origins of Mendelism* (1966), p. 18: "The notion of sexuality disturbed their orderly conception of an organic world in which insensitive, asexual plants formed the lower end of a 'chain of being' that ascended by imperceptible degrees through the different departments of the animal kingdom toward the perfection of man."

117
Linnaeus, "Le mariage des fleurs," cited by K. Hagberg, *Carl Linné, Le roi des fleurs* (1914), p. 46.

118
Trembley, *Instruction d'un père*, p. 110.

119
Siegesbeck, *Botanosophiae verioris Sciagraphia*, p. 49, cited in Stoever, *The Life of Sir Charles Linnaeus* . . . , translated by J. Trapp (London, 1794), p. 121. "I do not in all honesty know what to reply to these objections," says Browallius, "we are in effect constrained to silence by the very absurdity of the adversary." *Examen epicriseos siegesbeckianae in systema plantarum sexuale,*

Ch. Linnaei (Leyden, 1743), p. 19. Stillingfleet is ironical: "The world has now become so moderate that I have not yet heard that the Linnaean system is regarded as heretical even at the court of Rome." *Miscellaneous Tracts Relating to Natural History*, 2nd ed. (London, 1762), p. xxi, note c.

120
Smellie, *Philosophy of Natural History*, vol. I, sec. II, p. 247.

121
Alston, *A Dissertation on the Sexes of Plants*, pp. 314–15.

122
Bonnet, *L'Amour végétal ou les noces des plantes*, p. 32.

123
Ibid., pp. 55–56.

124
Goethe, *Œuvres d'histoire naturelle*, translated from the German (Paris and Geneva, 1837), pp. 321–22.

Chapter 4

1
Lamarck, *Histoire naturelle des animaux sans vertèbres*, 2nd ed. (Paris, 1835), introduction, p. 100. "In *inorganic bodies* and even in *vegetables*, the motions of the concrete parts of every kind are merely communicated.... In consequence of this, it has been found that the laws of these motions can be determined, giving rise to a special science that has been called *mechanics*." We have borrowed the expression "actions of plants" from Samuel Farr, *A Philosophical Enquiry into the Nature, Origin and Extent of Animal Motion* (London, 1771), p. 104: "If we consider very exactly the original Actions of mankind, we may find that these are the only motives by which sublunary beings are instigated to Action; and if we examine all the active natures in the universe, we may discover that the Actions of Plants too are of this kind."

2
Hooper, *Observations on the Structure and Economy of Plants, to Which Is Added the Analogy between the Animal and Vegetable Kingdom* (Oxford, 1797), p. 73. Someone may object that people were interested in this problem as early as the seventeenth century. An example might be Hooke, *Micrographia* (London, 1665), pp. 116–21, or Caspar Bose, *Dissertatio botanico-philosophica de motu plantarum sensus Aemulo* (Leipzig, 1728). In fact, attention was paid only to the motions of the sensitive plant, a curiosity. On this point, see Charles Webster, "The recognition of plant sensitivity by English botanists in the seventeenth century," *Isis* 57(1966):5–23.

3
A. von Haller, *Dissertation sur les parties irritables et sensibles des animaux,* French trans. (Lausanne, 1755), p. 5.

4
Dal Covolo, *A Discourse Concerning the Irritability of Some Flowers* (London, 1767), pp. 40–41.

5
G. Canguilhem, *La Connaissance de la vie,* 2nd ed. (Paris, 1971), p. 185.

6
Dal Covolo, *A Discourse,* p. 41: "Lastly, may it not be said, that the filaments of flowers are not muscles?"

7
Smith, "Observation sur l'irritabilité des végétaux," *Journal de physique,* Rozier 33(1788):49.

8
Ibid., p. 49.

9
Gmelin, *Irritabilitas vegetabilium, in singulis plantarum partibus explorata ulterioribusque experimentis confirmata* (Tübingen, 1768), p. 306.

10
Bonnet, "Contemplation de la nature," *Œuvres* (Neuchatel, 1781), vol. VIII, part X, p. 508.

11
Van Marum, *Dissertatio philosophica inauguralis de motu fluidorum in Plantis, experimentis et observationibus* (Groningen, 1773), p. 56. See also J. Carradori, "De l'irritabilité de laitron épineux," *Journal de physique,* Rozier 67(1808):405–13; Hedwig, *De fibrae vegetalis et animalis ortu* (Leipzig, 1789); Hope, *Quaedam de plantarum motibus et vita complectens* (Edinburgh, 1787).

12
Bonnet, "Contemplation de la nature," pp. 509–10.

13
Van Marum, "Seconde lettre à J. Ingen-Housz contenant quelques expériences et des considérations sur l'action des vaisseaux des plantes qui produit l'ascension et le mouvement de la sève," *Journal de physique,* Rozier 41(1792):216.

14
Bonnet, "Contemplation de la nature," p. 509. Bonnet suggested another direction of research: "Irritability resides in the gelatinous substance of the animal. Has the gelatinous substance of vegetables been studied carefully?" Ibid., p. 504.

15
Van Marum, "Seconde lettre à J. Ingen-Housz," p. 217.

16
Ibid., p. 218. Tiberius Cavallo carried out the same experiment: see *A Complete Treatise of Electricity in Theory and Practice* (London, 1777), p. 60. See also von Uslar, *Fragmente neuerer Pflanzenkunde* (Brunswick, 1794).

17
Coulon, *Dissertatio Academicae de mutata humorum in regno organico indolea vi vitali vasorum derivanda* (Leyden, 1789), pp. 12–3: "The irritability of vegetables, like that of animals, is sometimes completely destroyed by various substances: opium seems to be one of them. At Edinburgh, the sensitive plant was watered with decoctions of opium, and it lost its sensitivity." Other substances could also be used; see Tupper, *An Essay on the Probability of Sensation in Vegetables*, p. 47; see also von Humboldt, *Florae fribergensis specimen* (Berlin, 1793), pp. 167–68.

18
Brown, *Eléments de médecine*, French trans. (Paris, An VIII [1805]), chap. II, p. 3.

19
Girtanner, "Mémoire sur l'irritabilité considérée comme principe de vie dans la nature organisée," *Journal de physique*, Rozier 36(1790):425–26.

20
Ibid., p. 429.

21
Brown, *Eléments de médecine*, chap. I, p. 1.

22
E. Darwin, *Phytologia, or the Philosophy of Agriculture and Gardening* (London, 1800), sec. VIII, part 1, p. 135. For Erasmus Darwin, naturally, "the fibers and nerves which constitute these muscles are too fine to allow anatomical demonstration." Ibid., p. 132.

23
Ibid., p. 137. See also La Metherie, *Considérations sur les êtres organisés* (Paris, 1804), vol. I, p. 281: "They must therefore have organs of feeling, and we may assume that they have different external senses which transmit their sensations to them." See Bell, "The physiology of plants," in A. Hunter, *Georgical Essays* (1803), vol. I, Essay X, p. 542; and Bruce, "An account of the sensitive quality of the tree *Averrhoa carambola*," *Philosophical Transactions* no. 75 (1785):356–60.

24
Bonnet, "Contemplation de la nature," p. 417.

25
Tupper, *An Essay on the Probability of Sensation in Vegetables*, pp. 14–15 [cited passage is translated from French—trans.].

26
Percival, "Speculations on the Perceptive Power of Vegetables," in *Essays Medical, Philosophical and Experimental* (Warrington, 1789), vol. II, Essay III, p. 236 [cited passage is translated from French—trans.]. Percival is no doubt thinking of the *Dionaea muscipula* described by Ellis in *De Dionaea muscipula Planta irritabili nuper detecta ad . . . Perill. Car. a Linne Epistola* (Erlangen, 1771).

27
Tupper, *An Essay on the Probability of Sensation in Vegetables*, pp. 49–50. See also Percival, "Speculations," p. 6.

28
Hooper, *Observations on the Structure and Economy of Plants*, p. 73. On this point, see von Humboldt, *Expériences sur le Galvanisme*, French trans. (Paris, 1799), p. 250: "Do we have the right, however, to deny categorically that vegetables have nerves, merely because we have not yet discovered them? Was it not once asserted that worms have no nerves?" Hooper and von Humboldt made reference to the work of Sir Anthony Carlisle, "Observations upon the Structure and Œconomy of those Worms called Taeniae," *Transactions of the Linnaean Society* 2(1797):253–4.

29
Towson, "Objections against the perceptivity of plants," *Tracts and Observations on Natural History and Physiology* (London, 1799), p. 145.

30
Linnaeus, "La police de la nature," *L'Equilibre de la nature* (Paris, 1972), p. 118.

31
Bonnet, "Contemplation de la nature," p. 469.

32
G. Canguilhem, "Du singulier et de la singularité en épistémologie," *Etudes d'histoire et de philosophie des sciences*, (Paris, 1968), p. 215.

33
Tupper, *Probability of Sensation in Vegetables*, p. 24. See also Henri Home, *The Gentleman Farmer* (Edinburgh, 1776), p. 387: "All the trefoils may serve as a barometer to the husbandman as they always contract their leaves on an impending storm."

34
Linnaeus, *Philosophie botanique* (Paris, 1788), p. 320.

35
Hill, *Le Sommeil des plantes, et la cause du mouvement de la sensitive expliques par M. Hill* (Geneva and Paris, 1773), pp. 17–18.

36
Ibid., p. 13.

37
Bonnet, "Contemplation de la nature," p. 483. This explanation was proposed earlier by Dodart; see "Sur l'affectation de la perpendiculaire, remarquable dans toutes les tiges, dans plusieurs racines, et autant qu'il est possible, dans toutes les branches des plantes," *Mémoires de l'Académie des Sciences*, Paris (1700): 47–58.

38
Duhamel du Monceau, *La Physique des arbres* (Paris, 1758), vol. II, book IV, p. 140. The same explanation was proposed earlier by La Hire; see "Explication physique de la direction verticale et naturelle des tiges des plantes et des branches des arbres, et de leurs racines," *Mémoires de l'Académie des Sciences*, (1708): 231–33.

39
Ibid., p. 159. Dortous de Mairan had observed that the leaves of the sensitive plant continue to open even if the plant is kept in the dark: "M. de Mairan has observed that this phenomenon does not require that the plant be in sunlight or in the open air." "Observations botaniques," *Histoire de l'Académie Royale des Sciences*, Paris (1729):35. Duhamel du Monceau was in this instance merely adopting the results of earlier observers, from which, like Du Fay, he inferred "that the alternating motion of the sensitive depends neither on the temperature of the air nor on the light of day and darkness of night exclusively." "Observations sur la sensitive," *Mémoires de l'Académie des Sciences*, Paris (1736):90.

40
Duhamel du Monceau, "Observations," p. 155.

41
Linnaeus, "Somnus plantarum." *Amoenitates academicae* IV (1760):333–50.

42
Duhamel du Monceau, *La Physique des arbres*, vol. I, book IV, p. 172. See also Tournefort: "It is well to point out that by the word 'muscle' is meant a part composed of tissue consisting of fibres which are arranged in such a way that by their contraction they cause this same part to act necessarily in a determinate manner." "Observations physiques touchant les muscles de certaines

plantes," *Mémoires de l'Académie des Sciences*, Paris X (1688–1699):406.

43
Duhamel du Monceau, *La Physique des arbres*, p. 169.

44
Sénebier, *Mémoires physico-chimiques, sur l'influence de la lumière solaire pour modifier les êtres des trois règnes de la nature, et surtout du règne vegetal* (Geneva, 1782), vol. II, p. 294. On this point, see also Gleditsch, "Nouvelles expériences physiques sur l'accroissement et la diminution du mouvement extérieur des plantes par lequel les plantes s'écartent de leur direction perpendiculaire, suivant les diverses températures de l'air," *Mémoires de l'Académie des Sciences*, Berlin (1775):52–90.

45
A.-P. de Candolle, "Expériences relatives à l'influence de la lumière sur quelques végétaux," *Journal de physique*, Rozier 52(1801):128. On this point see M. Guédès, "Augustin-Pyramus de Candolle et les mouvements des végétaux," *Histoire et nature* (1974), new series, fasc. 2.

46
J. Hunter, *Traité sur le sang*, French trans. (Ostende, 1800), p. 93.

47
Duhamel du Monceau, *La Physique des arbres*, p. 141.

48
Knight, "On the direction of the radicle and germen during the vegetation of seeds," *Philosophical Transactions* 66(1806):100.

49
Lindsay, *An Inquiry into the Nature of Motion of the Sensitive, Sleeping and Moving Plants* (Jamaica: July 1790), p. 20 [passage cited trans. from French—trans.].

50
Whytt, *Essais physiologiques*, French trans. (Paris, 1759), p. 273. See also *An Essay on the Vital and Other Involuntary Motions of Animals*, 2nd ed. (Edinburgh, 1763), p. 289 n. 2: "The shrinking of the leaves of the sensitive plant when they are touched does not indicate any kind of feeling and in no way resembles the alternative contractions of irritated muscles."

51
Lamarck, *Inedits de Lamarck* (Paris, 1972): see the Discours d'ouverture de mon cours pour 1809 (Ms. 742, 2, pp. 20–21), p. 209. Another difference: animal irritability continues to manifest itself after death: "The animal keeps its irritability after death; the vegetable has none." Peschier, "Dissertation sur l'irritabilité des animaux et des plantes," *Journal de physique* 45(1794):356.

52
Lindsay, *An Inquiry*: see experiments 7, 14, and 23.

53
Ibid., pp. 22–3.

54
Lamarck, *Inédits de Lamarck*, p. 209. See also *Encyclopédie méthodique botanique* (Paris, 1783), vol. I, the article entitled "Acacia," pp. 17–18: "From this explanation, if it has any foundation, it emerges that the admirable phenomenon to which it applies is not due to any real sensation of the plant in which it is observed, as one might imagine, but rather that it is the effect of a purely mechanical cause."

55
Lindsay, *An Inquiry*, p. 24.

56
Broussonet, "Essai de comparaison entre les mouvements des animaux et ceux des plantes. Et description d'une espèce de sainfoin (*Hedysarum*) dont les feuilles sont dans un mouvement continuel," *Mémoires de l'Académie des Sciences*, Paris (1784):614.

57
Sénebier, *Physiologie végétale* (Geneva, 1799), vol. V, book III, p. 114.

58
Broussonet, "Essai de comparaison," p. 616.

59
Knight, "On the causes which influence the direction of growth of roots," *Philosophical Transactions* 71(1811):218.

60
Lamarck, *Histoire naturelle des animaux sans vertèbres*, introduction, p. 106.

61
A. von Haller, *Dissertation sur les parties irritables et sensibles des animaux* (1755), p. 82. Haller's reference to 1743 is thus an allusion to Boerhaave, *Praelectiones Academicae in proprias institutiones Medicae*, edidit et notas Albertus von Haller (Göttingen, 1743), vol. IV, p. 586, note a: "Furthermore, this entire theory is based on the following very simple phenomenon, which noone can deny: Any musculous fiber of a living animal, when irritated by any cause whatsoever, contracts continously in such a way that this latter character becomes one by which the imperfect animals can be distinguished from vegetables."

62
Scopoli, *Introductio ad Historiam naturalem* (Prague, 1777), p. 41.

63
A. von Haller, *Elements de physiologie de M. A. de Haller*, trans. from the Latin by Bordenave (Paris, 1769), part I, pp. 251–52.

64
Bonnet, "Contemplation de la nature," vol. VIII, p. 474.

65
Ibid., p. 460. Bonnet adds the further example of the reproductive function. "This is another of the relations that link the plant to the animal." Ibid., p. 517, n. 5.

66
Farr, *A Philosophical Enquiry*, p. 110.

67
Robinet, *De la nature*, vol. IV, p. 27.

68
Bonnet, "Contemplation de la nature," p. 486, n. 1. See also Camper, *Oratio de analogia inter animalia et stirpes* (Groningen, 1764).

69
Alphonse de Candolle, *Introduction à la botanique* (Paris, 1835), vol. I, p. 233.

70
A.-P. de Candolle and J.-B. Lamarck, *Flore française*, 3rd ed. (Paris, 1805), vol. I, p. 160.

71
Percival, *Perceptive Power of Vegetables*, p. 245.

72
J. Peschier, "Dissertation sur l'irritabilité des animaux et des plantes," *Journal de physique* 45(1794):356.

73
Tupper, *An Essay on the Probability of Sensation in Vegetables*, p. 66.

74
Duhamel du Monceau, *La Physique des arbres*, vol. I, p. 193. See also Tupper, *An Essay*, pp. 18–19: "There are plants and herbs which, like the porcupine and hedgehog, are provided with spines and stickers that protect them against the attack of many animals."

75
Ellis, *De dionaea muscipula planta irritabili nuper detecta ad . . . Car. a Linne epistola*, p. viii.

76
Smith, *An Introduction to Physiological and Systematical Botany* (London, 1807), p. 4.

77
Bonnet, "Contemplation de la nature," part X, p. 473.
78
Linnaeus, *Philosophie botanique*, aphorism 335, p. 328.
79
E. Ors, *Du Baroque* (Paris, 1968), p. 112. See also J. Rousset, *La Litterature de l'age baroque en France* (Paris, 1954), pp. 184–89.
80
Sénebier, *Physiologie vegetale*, vol. I, p. 17.
81
Bonnet, "Contemplation de la nature," chap. XXXIV, p. 510.

Conclusion

1
M. Foucault, *L'Ordre du discours* (Paris, 1971), p. 17.
2
On this point, see J. Schiller and T. Schiller, *Henri Dutrochet (Henri du Trochet, 1776–1847). Le Matérialisme mécaniste et la physiologie generale* (Paris: 1975).

Appendix

1
Author of the memoir entitled "Observations sur la structure et l'usage des principales parties des fleurs," *Mémoires de Académie des Sciences*, Paris (1711):210–30.

2
Antoine de Jussieu at first wrote and later crossed out: "My design is also not to challenge the observation of M. Geoffroy concerning the state of sterility he thought he saw in the pods of leguminous plants before the blossoming of the flower with the objection that I might well raise: that prior to the blossoming of the flowers, the dust from their stamens had already fallen onto the end of the pistil, since we find almost no dust on the tips of the stamens just after the flowers blossom."

3
Allusion to Tournefort's theory.

4
See his memoir, "Observations sur la végétation des truffes," *Mémoires de l'Académie des Sciences*, Paris (1711).

5
See his memoir, "Observations touchant la nature des plantes, et quelques-unes de leurs parties cachées ou inconnues," *Mémoires de l'Académie des Sciences*, Paris (1711).

6 Author of *L'Histoire physique de la mer* (Amsterdam, 1725).

7 Claude-Nicolas Billerer (1667–1759).

Bibliography

The bibliography is organized so as to correspond to the successive phases of the research. At the beginning I have listed historical works of a general nature, and at the end works in the history and philosophy of science and history of ideas whose scope is broader than that of the present study. There is a numbered section of the bibliography corresponding to each chapter of the book, and within each, two subsections, the first listing historic works, books and memoirs from the period under investigation, and the second listing critical books and articles of more recent date.

I. GENERAL AND SPECIALIZED WORKS IN THE HISTORY OF BOTANY

Bibliography of the Botanical Sciences

Boehmer, Georges R. *Systematisch-literarisches Handbuch der Naturgeschichte Oeconomie und anderer damit verwondten Wissenschaften und Künste.* 9 vols. Leipzig: J. F. Iunius, 1785–1789.

Haller, Albert von. *Bibliotheca botanica qua scripta ad rem herbarium facienta a rerum initiis recensentur.* 2 vols. Tiguri: Fuessli, 1771–72.

Jackson, Benjamin D. *Guide to the Literature of Botany, Being a Classified Selection of Botanical Works.* London: Longmans, Green, 1881.

Pritzel, George. *Thesaurus litteraturae botanicae amnium gentium inde a rerum botanicarum initiis ad nostra usque tempora quindecim millia operum recensens. Ed nova.* Leipzig: Brockhaus, 1872.

General History

Guyenot, Emile. *Les Sciences de la vie aux XVIIe et XVIIIe siècles.* Paris: Albin Michel, 1941.

Singer, Charles. *Histoire de la biologie.* Translated by Dr. Gibbon. Paris: Payot, 1934.

Singer, Charles. *Studies in the History and Method of Science.* vol. 2. Oxford: at the Clarendon Press, 1921.

History of Botany

Banal, J. A., Sr., and Abel Pilon. *Précis de l'histoire de la botanique pour servir de complément à l'étude du règne végétal, par L. G . . . , suivi d'un Appendice de géographie botanique avec Cartes.* 1869.

Combes, Raoul. *Histoire de la biologie végétale en France.* Paris: Alcan, 1933.

Davy de Virville. *Histoire de la botanique en France.* Paris: Société d'études d'enseignement supérieur, 1954.

Hoefer, Ferdinand. *Histoire de la botanique depuis les temps les plus recules jusqu'a nos jours.* Paris: Hachette, 1873.

Plantefol, Lucien. "Histoire de la botanique." *Troisième Centenaire de l'Académie des Sciences.* 2 vols. Paris: Gauthier-Villars, 1967.

Ritterbush, Philip C. *Overtures to Biology: The Speculations of Eighteenth-Century Naturalists.* New Haven, Conn.: Yale University Press, 1964.

Sachs, Julius von. *Geschichte der Botanik vom XVI. Jahrhundert bis 1860.* Munich: Oldenburg, 1876. Citations in the text are from the French translation by Henri de Varigny, *Histoire de la botanique du XVIe siècle à 1860.* Paris, 1892.

Sprengel, Kurt. *Geschichte der Botanik.* 2 vols. Leipzig: Brockhaus, 1817–18.

II. THE CENTRAL PROBLEMS OF BOTANY

Historic Works

ANTIQUITY

Aristotle. *Petits traités d'histoire naturelle.* Text established and translated by René Mugnier. Paris: 1965.

Aristotle. *The Basic Works of Aristotle.* Edited by Richard McKeon. New York: Random House, 1941.

Galen. *Œuvres choisies.* Translated into French by Charles Daremberg. 2 vols. Paris: 1854.

Hippocrates. *Œuvres complètes d'Hippocrate.* Translated into French and annotated by Emil Littré. 10 vols. Paris: 1839–1861.

RENAISSANCE

Bacon, Francis. *Histoire naturelle de M. Francis Bacon.* Paris: A. de Sommaville and A. Soubron, 1631.

Cardanus, Girolamo. *Les Livres intitulés de la subtilité.* French trans. Paris: G. le Noir, 1556.

Cesalpinus, Andraeus. *De plantis, Libri XVIe.* Florence: G. Marescotum, 1583.

Dordoens, Rembert. *Histoire des plantes.* Antwerp: J. Loe, 1557.

Duret, Claude. *Histoire admirable des plantes.* Paris: N. Buon, 1605.

Gerard, John. *The Herball, or General History of Plants.* London: Aslip, J. Norton, and K. Whitakers, 1633.

La Brosse, Guy de. *De la Nature, vertu et utilité des plantes, divisé en cinq livres.* Paris: R. Baragne, 1680.

Porta, Giambattista della. *Magiae Naturalis Libri Viginti.* Naples: 1589.

SEVENTEENTH AND EIGHTEENTH CENTURIES

Bayle, François. *Dissertationes physicae in quibus principia proprietarum in mistis, Oeconomia corporum in plantis et animalium.* The Hague: 1678.

Dodart, Denis. "Second mémoire sur la fécondité des plantes. Conjectures sur ce sujet." *Mémoires de l'Académie des Sciences,* Paris (1701):241–57.

Duhamel du Monceau, Henri-Louis. *De l'exploitation des bois.* 2 vols. Paris: H.-L. Guérin et L.-F. Delatour, 1764.

Feldmann, Bernhard. *Dissertatio physico-medica sistens comparationem plantarum et animalium.* Leyden, 1732.

Grew, Nehemiah. *The Anatomy of Vegetables, Begun, with a General Account of Vegetation Founded Thereon.* London: 1762.

Grew, Nehemiah. *Cosmologia Sacra, or a Discourse of the Universe as It Is the Creator and Kingdom of Good.* London: W. Rogers, S. Smith, and B. Walford, 1701.

Jussieu, Bernard de. "De la nécessité d'établir dans la méthode nouvelle des plantes une classe particulière pour les fougères, à laquelle doivent se rapporter, non seulement les champignons, les agarics, mais aussi les lichens." *Mémoires de l'Académie des Sciences*, Paris (1728).

La Mettrie, Julien Offray de. "Traité de l'âme," *Œuvres philosophiques*. London: 1751.

Lamy, Bernard. *Entretiens sur les sciences*. Critical edition, with an introduction by François Girbal and Pierre Clair. Paris: Presses Universitaires de France, 1966.

Necker, Noël Joseph de. *Phytozoologie philosophique*. Neuwied: Société typographique, 1790.

Ray, John. *L'Existence et la sagesse de Dieu manifestées dans les oeuvres de la création*. Utrecht: C. Broedelet, 1714.

Sénebier, Jean. *L'Art d'observer*. 2 vols. Geneva: J. J. Pachoud, 1775.

Swammerdam, Jean. *Histoire naturelle des insectes*. Dijon: Desventes, 1758.

Critical and Recent Works

Comte, Auguste. *Système de politique positive*. 4 vols. Paris: L. Mathias, 1851–1854.

Dagognet, François. *Des révolutions vertes*. Paris: Hermann, 1973.

Littré, Emile, and Robin, Charles. *Dictionnaire de médecine*. 10th ed. Paris: P. H. Nysten, 1855.

III. NUTRITION

Historic Works

Anonymous. "Extrait d'un discours sur la respiration des plantes prononcé à l'ouverture de l'Université dans le collège de Toulouse de la Compagnie de Jésus par le professeur de physique de la même Compagnie." *Mémoires pour l'histoire des sciences et des beauxarts*. Trevoux: 1712.

Bazin, Gilles Augustin. *Observations sur les plantes et leur analogie avec les insectes*. Strasbourg: Jean-Renaud Doulsecker, 1741.

Bertholon, abbé Pierre. *De l'électricité des végétaux*. Paris: P. F. Didot Jeune, 1783.

Birch, Thomas. *The History of the Royal Society of London*. 4 vols. London: A Millar, 1756–57.

Boerhaave, Hermann. *Elementa chimiae quae anniversario labore docuit in publicis, privatisque scholis.* Leyden: 1732; translated into French as *Eléments de chimie.* 2 vols. The Hague, 1748.

Bonnet, Charles. *Recherches sur l'usage des feuilles dans les plantes.* Göttingen and Leyden: E. Luzac, Jr., 1754.

Boyle, Robert. *The Works of the Honourable Robert Boyle, to Which is Prefixed the Life of the Author.* Edited by T. Birch. 5 vols. London: A Millar, 1744.

Bradley, Richard. "Observations and experiments relating to the motion of the sap in vegetables." *Philosophical Transactions* 349 (1716):486–90.

Bradley, Richard. *A Philosophical Treatise of Agriculture by G. A. Agricola, the Whole Revised and Compared with the Original.* London: 1721.

Bradley, Richard. *The Gentleman and Farmer's Kalendar, Directing What Is Necessary to Be Done Every Month.* London: 1718.

Bucquet, Jean-Baptiste-Michel. *Introduction à l'étude des corps naturels, tirés du règne végétal.* 2 vols. Paris: Vve Herissant, 1773.

Buffon, Georges Louis Leclerc, Comte de. *Histoire naturelle, générale et particulière avec la description du Cabinet du Roy.* 45 vols. Paris: Imprimerie Royale, 1749–1804.

Dedu, Nicolas. *De l'âme des plantes, de leur naissance, de leur nourriture et de leur progrès. Essai de physique.* Paris: E. Michallet, 1682.

Descartes, René. "Les principes de la philosophie." *Œuvres et lettres.* Edited by Andre Bridoux. Paris: 1953.

Diderot, Denis. Article entitled "Animal" in *L'Encyclopédie.* 1781.

Duhamel du Monceau, Henri-Louis. "Recherches physiques sur la cause du prompt accroissement des plantes dans les temps de pluies. Et plusieurs observations à ce sujet." *Mémoires de l'Académie des Sciences,* Paris (1729):349–60.

Duhamel du Monceau, Henri-Louis. "Anatomie de la poire." *Mémoires de l'Académie des Sciences,* Paris (1730):299–324.

Duhamel du Monceau, Henri-Louis. Sur le développement et la crue des os des animaux." *Mémoires de l'Académie des Sciences,* Paris (1742):354–70.

Duhamel du Monceau, Henri-Louis. "Sur la réunion des plaies des arbres et des animaux; et sur les greffes ou incisions, tant végétales qu'animales." *Mémoires de l'Académie des Sciences,* Paris (1746):319–58.

Duhamel du Monceau, Henri-Louis. "Recherches sur la formation des couches ligneuses dans les arbres." *Mémoires de l'Académie des Sciences*, Paris (1751):23–35.

Duhamel du Monceau, Henri-Louis. *Eléments d'agriculture.* 2 vols. Paris: H.-L. Guérin and L.-F. Delatour, 1762.

Fouquet, Henri. Article entitled "Sécrétion" in *L'Encyclopédie*, 1781.

Geoffroy, Etienne François. *Thèse soutenue aux écoles.* Paris: L. d'Houry, 1705.

Gersten, Christian Ludwig. *Tentamina systematis novi ad mutationes barometri ex natura elateris aerei demonstrandas, cui adjecta sub finem dissertatio Roris decidui errorem antiquum et vulgarem per observationes et experimenta nova excutiens.* Frankfurt: P. Varentrapp, 1733.

Gesner, Johannis. *Dissertationes physicae de vegetabilius quarum prior partium vegetationis structuram.* Leyden, 1743.

Gouan, Antoine. *Discours sur la cause du mouvement de la sève dans les plantes.* Montpellier: G. Izar and A. Ricard: 1803.

Grew, Nehemiah. *The Anatomy of Plants, with an Idea of a Philosophical History of Plants. And Several other Lectures Read before the Royal Society.* London: W. Rawlins, 1682.

Guettard, Jean Etienne. "Mémoire sur les corps glanduleux des plantes, leurs filets ou poils, et les matières qui suintent des unes ou des autres." *Mémoires de l'Académie des Sciences*, Paris (1745):261–306.

Guettard, Jean Etienne. *Observations sur les plantes.* 2 vols. Paris: Durand, 1747.

Guettard, Jean Etienne. "Mémoire sur la transpiration insensible des plantes." *Mémoires de l'Académie des Sciences*, Paris (1748):569–92.

Guettard, Jean Etienne. "Second mémoire sur la transpiration insensible des plantes." *Mémoires de l'Académie des Sciences*, Paris (1749):265–317.

Hales, Stephen. *Vegetable Staticks, or an Account of Some Statical Experiments on the Sap in Vegetables.* London: W. and J. Innys, 1727.

Hales, Stephen. *Statical Essays, Containing Haemastaticks.* London: W. Innys and R. Mayby, 1733.

Hartsoeker, Nicolas. *Eclaircissements sur les conjectures physiques.* 2 vols. Amsterdam: P. Humbert, 1710.

Helmont, Jean-Baptiste van. *Les Œuvres de Jean-Baptiste van Helmont traitant des principes de médecine et physique pour la guérison assurée des maladies.* From the translation by M. Jean Le Conte. Lyon: J. A. Huguetan and G. Barbier, 1670.

Hertel, Johann Gottlob. *Dissertationem physiquam de plantarum transpirationem.* Leipzig: J. C. Langenheim, 1735.

Hill, John. *The Vegetable System, or the Internal Structure and the Life of Plants.* 2nd ed. London: R. Baldwin, 1773.

Hoffman, Frederick. *La Médecine raisonné.* Translated from the Latin by J. Bruhier. 9 vols. Paris: 1739–1749.

Huygens, Christian. "Le Cosmotheros." *Œuvres complètes de Christian Huygens.* 26 vols. The Hague: 1944.

Jaucourt, Chevalier de. Article "Plant" in the *Encyclopédie*, 1781.

Jussieu, Antoine Laurent de. *An œconomiam animalem inter et vegetalem analogia?* Paris; 1770.

Kant, Immanuel. *Rêve d'un visionnaire.* Translated and with an introduction by Francis Courtex. Paris: Vrin, 1967.

La Hire, Philippe de. "Sur la cause de l'élévation du suc nourricier dans les plantes." *Histoire de l'Académie Royale des Sciences*, Paris 2 (1686):185–86.

La Hire, Philippe de. "Expériences servant d'eclaircissement à l'élévation du suc nourricier dans les plantes." *Mémoires de l'Académie des Sciences*, Paris X (1666–1699):317–19.

La Metherie, Jean-Claude de. *Vues physiologiques sur l'organisation animale et végétale.* Amsterdam and Paris: P.-F. Didot Jeune, 1780.

La Mettrie, Julien Offroy de. "L'Homme-plante." *Œuvres philosophiques.* London: 1751.

La Quintinie, Jean de. *Instruction pour les jardins fruitiers et potagers avec un traité des orangers, et des réflexions sur l'agriculture. Nouvelle édition revue, corrigée et augmentée d'une instruction pour la culture des fleurs.* 2 vols. Paris, 1756.

Leeuwenhoek, Anthony van. "An abstract of a letter from M. Leeuwenhoek concerning the appearances of several woods, and their vessels." *Philosophical Transactions* 148(1683):197–208.

Magnol, Pierre. "Sur la circulation de la sève dans les plantes." *Histoire de l'Académie Royale des Sciences*, Paris (1709):44–49.

Major, Jean Daniel. *Dissertation botanica de planta monstrosa Gottorpiensi.* Schleswig: J. Carstens, 1665.

Malpighi, Marcello. *Opera omnia.* London: R. Littlebury, 1687.

Mariotte, abbé Edme. "Premier essai sur la végétation des plantes." Œuvres. 2 vols. The Hague: J. Neaulme, 1740.

Mayow, Johannes. *Opera omnia medico-physica, tractatibus quinque comprehensa.* The Hague: A. Leers, 1681.

Montesquieu, Charles Louis de. "Observations sur divers sujets d'histoire naturelle." *Œuvres de Montesquieu.* Paris: E.-A. Lequieu, 1819.

Musschenbroek, Pieter van. *Cours de physique expérimentale et mathématique.* Translated by M. Sigaud de la Fond. Leyden: 1769.

Mustel, Nicolas Alexandre. "New observations upon vegetation." *Philosophical Transactions* 63(1773):126–36.

Mustel, Nicolas Alexandre. *Traité théorique et pratique de la végétation.* 4 vols. Paris and Rouen: Le Boucher le Jeune, 1781–1784.

Nieuwentyt, Bernard. *L'Existence de Dieu demontrée par les merveilles de la nature.* Amsterdam and Leipzig: Arkstee and Merkus, 1760.

Parent, Antoine. "Sur la nourriture des plantes." *Histoire de l'Académie des Sciences,* Paris (1711):42–50.

Parsons, James. *Philosophical Observations on the Analogy between the Propagation of Animals and Vegetables.* London: C. Davis, 1752.

Perrault, Claude. "De la circulation de la sève dans les plantes." *Essais de physique.* 4 vols. Paris: J. B. Coignard, 1680–1688.

Reichel, George Christian. *De vasis plantarum spiralibus.* Leipzig: Breitkopfia, 1758.

Renéaume, René-Louis. "Observations sur le suc nourricier des plantes." *Mémoires de l'Académie des Sciences,* Paris (1707): 276–86.

Sarrabat, Nicolas (pseudonym "de la Baïsse"). *Dissertation sur la circulation de la sève dans les plantes.* Bordeaux: P. Brun, 1733.

Saussure, Horace Benedict de. *Observations sur l'écorce des des feuilles et des pétales.* Geneva: 1762.

Tull, Jethro. *The Horse Hoeing Husbandry: or an Essay on the Principles of Tillage and Vegetation.* London: G. Strahan, 1733–1736.

Vicq d'Azir, Felix. *Œuvre de Vicq d'Azir.* 6 vols. Paris: L. Duprat-Duverger, 1805.

Walker, John. "Experiments on the motion of the sap in trees." *Transactions of the Royal Society of Edinburgh* 1 (1788):3–40.

Wolf, Christian. *Vernünftige Gedanken von den Wirkungen der Natur*. Hall: 1723.

Woodward, John. "Some thoughts and experiments concerning vegetation." *Philosophical Transactions* 253(1699):193–227.

Critical and Recent Works

Cuvier, Georges, and Magdelaine de Saint-Agy. *Histoire des sciences naturelles depuis leur origine jusqu'à nos jours chez tous les peuples connus*. 5 vols. Paris: Masson, 1841–1845.

Flourens, Pierre. *Histoire des travaux et des idées de Buffon*. 3rd ed. Paris: Garnier Frères, 1870.

Hanks, Leslie. *Buffon avant "l'histoire naturelle."* Paris: Presses Universitaires de France, 1966.

Patterson, T. S. "John Mayow in contemporary setting. A contribution to the history of respiration and combustion." *Isis* 15, no. 42 (1931).

Plantefol, Lucien. "Duhamel du Monceau." *Dix-huitième siècle* 1 (1969):123–37.

IV. GENERATION

Historic Works

Adanson, Michel. *Famille des plantes*. 2 vols. Paris: Vincent, 1763.

Adanson, Michel. "Examen de la question, si les espèces changent parmi les plantes: nouvelles expériences tentées à ce sujet." *Mémoires de l'Académie des Sciences*, Paris (1769):31–48.

Alston, Charles. *A Dissertation on Botany*. Translated from the Latin. London: B. Dod, 1754.

Alston, Charles. "A dissertation on the sexes of plants." *Essays and Observations, Physical and Literary*. Read before the Philosophical Society in Edinburgh, 1771, pp. 228–318.

Bachelot de la Pylaie, Auguste-Jean-Marie. *Études cryptogamiques ou monographies de divers genres de mousses*. Paris: 1815.

Blair, Patrick. *Botanick Essays*. London: W. and J. Innys, 1720.

Bondary, Auguste Fougeroux de. "Mémoire sur la fécondation des plantes." *Journal de physique*, Rozier (1775):23–30.

Bonnet, Charles. "Idées sur la fécondation des plantes." Œuvres d'histoire naturelle et de philosophie de Charles Bonnet. Neuchâtel: S. Fauche, 1781, vol. X.

Bonnet, Charles. "Lettres." Ibid., vol. XII.

Bonnet, Charles. L'Amour végétal ou les noces des plantes, augmentée des lettres de Jean-Jacques Rousseau sur la botanique. 2nd ed. Paris: Maugeret fils, 1809.

Bradley, Richard. New Improvements of Planting and Gardening, both Philosophical and Practical, Explaining the Motion of the Sap and the Generation of Plants. 3 vols. London: W. Mears, 1717–18.

Bradley, Richard. A Philosophical Account of the Works of Nature. London: W. Mears, 1721.

Browallius, Johannes. Examen epicriseos Siegesbeckianae in systema plantarum sexuale. Leyden: 1743.

Bulliard, Pierre. Histoire des champignons de la France ou Traité élémentaire renfermant dans un ordre méthodique les descriptions et les figures des champignons qui croissent naturellement en France. 2 vols. Paris: The author (and Leblanc), 1791–1812.

Burckhard, Johann Heinrich. Epistola ad illustrem . . . dominum Godofredum Guillelmum Leibnitium, . . . qua characterem plantarum naturalem nec a radicibus, nec ab aliis plantarum partibus . . . Helmstedt: J. Drimborni, 1701.

Camerarius, Rudolf. "Semina mori subentanea." Ephemerides Academicae naturae curiosorum. Frankfurt: 1691, pp. 212–3.

Camerarius, Rudolph. De sexu plantarum epistola. Tübingen: 1694.

Candolle, Augustin-Pyramus de. Physiologie végétale ou exposition des forces et des fonctions vitales des végétaux. 3 vols. Paris: Bechet Jeune, 1832.

Condorcet, Antoine Nicolas, Marquis de. Eloges des Académiciens de l'Académie Royale des Sciences. 5 vols. Paris: 1799.

Correa da Serra. "On the fructification of the submersed algae." Philosophical Transactions 86 (1696):494–505.

Darwin, Erasmus. Zoonomia, or the Laws of Organic Life. 3rd ed., corrected. 3 vols. London: J. Johnson, 1801.

Desfontaines, René-Louis. "Observations sur l'irritabilité des organes sexuels d'un grand nombre de plantes." Mémoires de l'Académie des Sciences, Paris (1787):468–80.

Dudley, Paul. "Observations on some of the plants in New England with remarkable instances of the nature and power of vegetation." *Philosophical Transactions* 385(1724):194–200.

Fontenelle, Bernard Le Bovier de. "Sur la génération de l'homme par des oeufs." *Histoire de l'Académie des Sciences*, Paris (1701):38–45.

Gaertner, Joseph. *De Fructibus et seminibus plantarum*. 3 vols. Stuttgart: The author, 1788–1807.

Gardeen, Georges. "A discourse concerning the modern theory of generation." *Philosophical Transactions* 192 (1691):474–83.

Geoffroy, Claude. "Observations sur la structure et l'usage des principales parties des fleurs." *Mémoires de l'Académie des Sciences*, Paris (1711):210–30.

Gleditsch, Johann Gottlieb. "Essai d'une fécondation artificielle faite sur l'espèce de palmier qu'on nomme Palma dactylifera folio flabelliformi." *Memoirs of the Academy of Sciences*, Berlin (1749):103–8.

Gleditsch, Johann Gottlieb. "Dissertation sur un pommier à tige basse, en buisson d'une espèce dégénérée, femelle, apétale, et de ses variétés." *Memoirs of the Academy of Sciences*, Berlin (1754):76–91.

Gleditsch, Johann Gottlieb. "Exposition d'une fécondation artificielle des huîtres et des saumons, qui est appuyée sur des expériences certaines, faite par un habile Naturaliste." *Memoirs of the Academy of Sciences*, Berlin (1764):47–64.

Gleditsch, Johann Gottlieb. "Relation de la fécondation artificielle d'un palmier femelle, réitérée pour la troisième fois, et avec un plein succes, dans le jardin botanique de l'Académie Royale de Berlin." *Memoirs of the Academy of Sciences*, Berlin (1767):3–19.

Gleichen, Frederick William, known as Russworm. *Découvertes les plus nouvelles dans le règne végétal ou observations microscopiques sur les parties de la génération des plantes*. Translated from the German. Nuremberg: Vve. de C. De Launoy, 1770.

Gleichen, Frederick William. *Dissertation sur la génération, les animalcules spermatiques, et ceux d'infusion, avec des observations microscopiques sur le sperme, et les différentes infusions*. Translated from the German. Paris: Debure, 1800.

Goethe, Johann Wolfgang von. *Œuvres d'histoire naturelle de Goethe, comprenant divers mémoires d'anatomie comparée, de botanique et de géologie*. Translated and annotated by C. F. Martins. Paris: A. Cherbuliez, 1837.

Guettard, Jean Etienne. "Observations par lesquelles on détermine le caractère générique de la plante appelée Marsilea, plus exactement qu'il ne l'a été jusqu'à présent." *Mémoires de l'Académie des Sciences*, Paris (1742):547–55.

Hartsoeker, Nicolas. *Essay de Dioptrique*. Paris: J. Anisson, 1694.

Hasselquist, Frederick. *Voyage dans le Levant, dans les années 1749, 1750, 1751, 1752*. French translation. Paris: Delalaian, 1769.

Hedwig, Johann. *Theoria generationis et fructificationis plantarum cryptogamicarum Linnaei*. Petersburg Academiae imp. scientarum, 1784.

Henschel, August Wilhelm. *Von der Sexualität des Pflanzen*. Breslow: 1820.

Jussieu, Antoine de. *Du rapport des plantes avec les animaux tiré de la différence de leurs sexes*. Ms. 1260, Bibliothèque du Museum, 1721.

Jussieu, Antoine Laurent, and de Mirbel. "Rapport sur le mémoire de M. Desvaux sur les lycopodiacées, et monographie de cette famille." *Journal de physique* 76(1813):320–31.

Jussieu, Bernard de. "Histoire d'une plante connue par les botanistes sous le nom de Pilularia." *Mémoires de l'Académie des Sciences*, Paris (1740):263–74.

Kastner, Karl. "Anmerkungen über die muthmasslichen Gedanken von dem Staube der Pflanzen." *Hamburgisches Magazin* 3(1752):11–24.

Knight, Thomas Andrew. "An account of some experiments on the fecundation of vegetables." *Philosophical Transactions* 89(1799):195–204.

Koelreuter, Josef Gottlieb. *Vorläufige Nachricht von einigen das Geschlecht der Pflanzen betreffenden Versuchen und Beobachtungen*. Leipzig: 1761–66.

Lamouroux, Jean Felix Vincent. *Dissertation sur plusieurs espèces de Fucus*. Agen: R. Noubel, 1805.

Leibniz, Gottfried Wilhelm. "Nouveaux essais sur l'entendement humain." *Œuvres philosophiques*. Amsterdam and Leipzig: 1765.

Linnaeus, Carolus. *Systema naturae*. Leyden: 1735.

Linnaeus, Carolus. *De nuptiis et sexu plantarum*. Ed. Afzelius, 1828.

Linnaeus, Carolus. "Sponsalia plantarum." *Amoenitates academicae*. Leyden: 1749. vol. 1.

Linnaeus, Carolus. "Dissertation sur les sexes des plantes." *Journal de physique*, Rozier 32(1788):440–62.

Linnaeus, Carolus. "Economie de la nature." *Equilibre de la nature.* Paris: Vrin, 1972.

Logan, James. *Experiments and Considerations of Generation of Plants.* London: C. Davis, 1747.

Ludwig, Christian Gottlieb. *Dissertatio de sexu plantarum.* Leipzig, 1737.

Marti, Antonio de. *Experimentos y observaciones sobre los sexos y fecundacion de las plantas.* Barcelona: la viuda Piferrer, 1791.

Maupertuis, Pierre Louis Moreau de. "Venus physique." *Œuvres.* Lyon, 1768, vol. 2.

Miller, Philip. *The Gardeners and Florists Dictionary, or a Complete System of Horticulture.* 2 vols. London: C. Rivingston, 1724.

Miller, Philip. "Letter to Mr. Bradley, October 6, 1721." In R. Bradley, *A General Treatise of Husbandry and Gardening.* 3 vols. London: T. Woodward, 1724.

Möller, Georg Friedrich. "Fortsetzung der muthmasslichen Gedanken vom Bluhmenstaube." *Hamburgisches Magazin*, Hamburg and Leipzig III (1752):410–55.

Möller, Georg Friedrich. "Muthmassliche Gedanken, von dem Staube der Pflanzen während der Bluthe." *Hamburgisches Magazin*, Hamburg and Leipzig III (1748):454–76.

Morland, Samuel. "Some new observations upon the part and use of flower in plants." *Philosophical Transactions* 287(1703):1474–78.

Necker, Noël Joseph de. *Physiologie des corps organises.* Bouillon: aux dépens de la Société typographique, 1775.

Necker, Noël Joseph de. "Eclaircissements sur la propagation des Filicées en général." *Mémoires de l'Académie des Sciences*, Mannheim, 3(1775):275–318.

Needham, John Tuberville. *Nouvelles découvertes faites avec le microscope.* Leyden: 1747.

Palisot de Veauvois, Ambroise Marie Francois. *Prodrome des cinquième et sixième familles de l'Aetheogamie, les mousses et les lycopodes.* Paris: Fournier, 1805.

Pontedera, Giulio. *Anthologia, sive de Floris natura libri tres.* Pavia: J. Maufre, 1720.

Pultney, Richard. *Historical and Biographical Sketches of the Progress of Botany in England, from Its Origin to the Linnaean System.* 2 vols. London: T. Cadell, 1790.

Ray, John. *Stirpium Europearum extra Britannias nascentium silloge*. London: S. Smith and B. Walford, 1694.

Réaumur, Antoine Ferchault de. "Description des fleurs et des graines de divers fucus, et quelques autres observations physiques sur ces mêmes plantes." *Mémoires de l'Académie des Sciences*, Paris (1711):282–302.

Reynier, Louis. "Résultats de quelques expériences relatives à la génération des plantes." *Journal de physique*, Rozier 31 (1787): 321–28.

Rotheram, John. *The sexes of plants vindicated: in a letter to Mr. William Smellie*. Edinburgh: W. Creech, 1790.

Rousseau, Jean-Jacques. *Œuvres complètes de J.-J. Rousseau*. 37 vols. Paris: Berlin, 1793.

Schelver, Franz Joseph. *Kritik der Lehre von den Geschlechtern der Pflanzen*. Heidelberg: 1812–1823.

Siegesbeck, Johann Georg. *Botanosophia verioris brevis sciagraphia in usum discentium adornata; accedit ob argumenti analogiam epicrisis in clar. Linnaei nuperrime evulgatum systema plantarum sexuale, et huic superstructam methodum botanicum*. Petersburg: 1737.

Smellie, William. *The Philosophy of Natural History*. 2 vols. Edinburgh: W. Creech, 1790–1799.

Spallanzani, Lazarro. *Expériences pour servir à l'histoire de la génération des animaux et des plantes*. 3 vols. Pavia, 1787.

Sprengel, Christian Conrad. *Das entdeckte Geheimnis der Natur im Bau und in der Befruchtung der Blumen*. Berlin: F. Vieweg der aelter, 1793.

Stillingfleet, Benjamin. *Miscellaneous Tracts Relating to Natural History*, 2nd ed. London: R. and J. Dodsley, 1762.

Tournefort, Joseph Pitton de. "Introduction à la botanique." in *Tournefort*. Paris: Muséum national d'histoire naturelle, 1957.

Trembley, Abraham. *Instructions d'un père à ses enfants, sur la nature et sur la religion*. 2 vols. Geneva: J. S. Cailler, 1775.

Vaillant, Sebastien. *Discours sur la structure des fleurs, leurs différences et l'usage de leurs parties*. Leyden: P. Vander Aa, 1718.

Volta, Giorgio Serafino. "Nuove ricerche ed osservazioni sopra il sessualismo di alcune Piante." *Memorie della Academia di Scienza Bella Lettere ed Arti*, Mantova 1: 225–68.

Critical and Recent Works

Beer, Sir Gavin de. "Jean-Jacques Rousseau: Botanist." *Annals of Science* 10 (1954):189–223.

Bonnier, Gaston. *Le Monde végétal*. Paris: Flammarion, 1913.

Olby, Robert C. *Origins of Mendelism*. London: Constable, 1966.

Zirkle, Conrad. "Animals impregnated by the wind." *Isis* 25 (1936):95–130.

V. MOVEMENT

Historic Works

Bell, George. "The Physiology of Plants." In Alexander Hunter, *Georgical Essays*. York: Wilson and Spence, 1803, vol. 1, Essay X, pp. 519–43.

Boerhaave, Hermann. *Praelectiones academicae in proprias institutiones rei medicae*. Edited and annotated by Albert von Haller. Göttingen: 1743.

Bonnet, Charles. "Contemplation de la nature." *Œuvres d'histoire naturelle et de philosophie*. Neuchatel: S. Fauche, 1781, vol. 3.

Bose, Caspar. *Dissertatio botanico-philosophica de motu plantarum sensus aemulo*. Leipzig: Breitkopf, 1728.

Broussonet, Pierre Marie Auguste. "Essai de comparaison entre les mouvements des animaux et ceux des plantes. Et description d'une espèce de sainfoin (Hedysarum) dont les feuilles sont dans un mouvement continuel," *Mémoires de l'Académie des Sciences*, Paris (1784):609–21.

Brown, John. *Elementa medicinae*. Mediolani: excudebat J. Galealius, 1792; translated into French as *Eléments de médecine*. Paris: Domonville and Gabon, 1801.

Bruce, Robert. "An Account of the Sensitive Quality of the Three Averrhoa carambola." *Philosophical Transactions* 75 (1785):356–60.

Camper, Pieter. *Oratio de analogia inter animalia et stirpes*. Gröningen: Spandau, 1764.

Candolle, Augustin-Pyrame de. "Expériences relatives à l'influence de la lumière sur quelques végétaux." *Journal de physique*, Rozier 52 (1801):124–30.

Candolle, Alphonse de. *Introduction à l'étude de la botanique, ou Traité élémentaire de cette science*. 2 vols. Paris: Roret, 1835.

Carradori, Giovacchino. "De l'irritabilité du laitron épineux." *Journal de physique*, Rozier 67(1801):405–13.

Cavallo, Tiberius. *A Complete Treatise of Electricity in Theory and Practice; with Original Experiments*. London: E. and C. Dilly, 1777.

Coulon, Julius Vitringa. *Dissertatio academica de mutata humorum in regno organico indolea vi vitali vasorum derivanda*. Leyden: Abraham and Jan Honkoop, 1789.

Covolo, Giambattista, conte dal. *Discorso della irritabilità d'alcuni fiori*. Firenze, 1764; translated into English as *A Discourse Concerning the Irritability of Some Flowers*. London: J. Dodsley, S. Baker and G. Leigh, and T. Payne, 1767.

Darwin, Erasmus. *Phytologia; or the Philosophy of Agriculture and Gardening*. London: J. Johnson, 1800.

Dodart, Denis. "Sur l'affectation de la perpendiculaire, remarquable dans toutes les tiges, dans plusieurs racines, et autant qu'il est possible, dans toutes les branches des plantes. *Mémoires de l'Académie des Sciences*, Paris (1700):47–58.

Du Fay, Charles François de Cysternay. "Observations sur la sensitive." *Mémoires de l'Académie des Sciences*, Paris (1736):87–110.

Duhamel du Monceau, Henri-Louis. *La Physique des arbres, où il est traité de l'anatomie des plantes et de l'économie végétale*. 2 vols. Paris: H.-L. Guérin et L.-F. Delatour, 1758.

Ellis, John. *De dionaea muscipula planta irritabili nuper detecta ad . . . Car. a Linne epistola; Beschreibung der Dionaea muscipula . . . aus dem Englischen übersetzt und herausgegeben von D. Johann Christian Daniel Schreber*. Erlangen: W. Walther, 1771.

Farr, Samuel. *A Philosophical Enquiry in the Nature, Origin and Extent of Animal Motion*. London: T. Becket, 1771.

Girtanner, Christophe. "Mémoire sur l'irritabilité, considerée comme principe de vie dans la nature organisée." *Journal de physique* Rozier 36(1790):422–40.

Gleditsch, Johann. "Nouvelles expériences physiques sur l'accroissement et la diminution du mouvement extérieur des plantes par lequel les plantes s'écartent de leur direction perpendiculaire, suivant les diverses températures de l'air." *Memoirs of the Academy of Sciences*, Berlin (1665):52–90.

Gmelin, Johann. *Irritabilitas vegetabilium, in singulis plantarum partibus explorata ulterioribusque experimentis confirmata*. Tübingen: Sigismund, 1768.

Haller, Albrecht von. *Dissertation sur les parties irritables et sensibles des animaux*. Translated from the Latin by M. Tissot. Lausanne: Marc-Michel Bousquet, 1755.

Haller, Albrecht von. *Eléments de physiologie de M. Alb. de Haller*. Translated from the Latin by Bordenave. Paris: Guillyn, 1769.

Hedwig, Johann. *De fibrae vegetalis et animalis ortu*. Leipzig: Kindelia, 1789.

Hill, John. *The Sleep of Plants, and Cause of Motion in the Sensitive Plant, Explain'd in a Letter to C. Linnaeus, Professor of Botany at Upsal*. London: R. Baldwin, 1757.

Home, Henri Lord Kames. *Gentleman Farmer*. Edinburgh: W. Creech, 1776.

Hooke, Robert. *Micrographia, or Some Physiological Descriptions of Minute Bodies Made by Magnifying Glasses*. London: J. Martyn and J. Allestry, 1665.

Hooper, Robert. *Observations on the Structure and Economy of Plants; to Which Is Added the Analogy between the Animal and Vegetable Kingdom*. Oxford: Fletcher and Co., and Rivington and Murray and Highley of London, 1797.

Hope, Thomas Charles. *Tentamen inaugurale quaedam de plantarum motibus et vita complectens*. Edinburgh: Balfour and Smellie, 1787.

Humboldt, Frederick Alexander von. *Aphorismi ex doctrina phyiologiae chimicae plantarum, Impr. ejus Flora Fribergensis*. Berlin: H. A. Rottmann, 1793.

Humboldt, Frederick Alexander von. *Experiences sur le galvanisme et en général sur l'irritation des fibres musculaires et nerveuses*. French trans. Paris: J.-F. Fuchs, 1799.

Hunter, John. *A Treatise on the Blood, Inflammation, and Gunshot Wounds, by the Late John Hunter*. London: George Nicol, 1794.

Knight, Thomas, Andres. "On the direction of the radicule and germen during the vegetation of seeds." *Philosophical Transactions* 66(1806):99–108.

Knight, Thomas Andrew, "On the causes which influence the direction of growth of roots." *Philosophical Transactions* 71(1811):209–19.

La Hire, Philippe de. "Explication physique de la direction verticale et naturelle des tiges des plantes et des branches des arbres, et de leurs racines." *Mémoires de l'Académie des Sciences*, Paris (1708):158–9.

Lamarck, Jean Baptiste de. *Encyclopédie méthodique botanique.* 13 vols. Paris: Panckoucke, 1783–1817.

Lamarck, Jean Baptiste de. *Inédits de Lamarck.* With an introduction by M. Vachon, G. Rousseau, and Y. Laissus. Paris: Masson and Cie, 1972.

Lamarck, Jean Baptiste de. *Historie naturelle des animaux sans vertèbres.* 2nd ed. 11 vols. Paris: J.-B. Baillière, 1835.

Lamarck, Jean Baptiste de. *Candolle, A.-P. de. Flore Française.* 3rd ed. 3 vols. Paris: H. Agasse, 1805.

La Metherie, Jean Claude de la. *Consideration sur les êtres organisés.* 2 vols. Paris: Courcier, An VIII [1805].

Lindsay, John. "An inquiry into the nature of the motion of the sensitive, sleeping and moving plants." Jamaica: July 1790. *Letters and Papers of the Royal Society* 89 (decade IX, no, 199).

Linnaeus, Carolus. "Somnus plantarum." *Amoenitates academicae*, Leyden IV (1760):333–50.

Linnaeus, Carolus. "La police de la nature." *L'Equilibre de la nature.* Paris: Vrin, 1972.

Linnaeus, Carolus. *Philosophie botanique.* Translated from the Latin. Paris, Cailleau, and Rouen: Le Boucher, 1788.

Mairan, Jean-Jacques Dortous de. "Observations botaniques." *Histoire de l'Académie Royale de Sciences*, Paris (1729):35.

Marum, Martinus van. *Dissertatio philosophica de motu fluidorum in plantis, experimentis et observationibus indigato.* Gronigen: 1773.

Marum, Martinus van. "Seconde lettre de M. van Marum à Jean Ingen-Housz contenant quelques expériences et des considérations sur l'action des vaisseaux des plantes qui produit l'ascension et le mouvement de leur sève." *Journal de physique*, Rozier 41(1792):214–20.

Percival, Thomas. "Spéculations on the perceptive Power of Vegetables." *Essays Medical, Philosophical and Experimental.* 4th ed. 2 vols. Warrington: W. Eyres, 1788–89.

Peschier, Jacques-Louis. "Dissertation sur l'irritabilité des animaux et des plantes." *Journal de physique* 45(1794):343–57.

Robinet, Jean Baptiste René. *De la nature.* 5 vols. Amsterdam: E. van Harrevelt, 1786.

Scopoli, Giovanni Antonio. *Introductio ad Historiam naturalem,* sistens genera lapidium, plantarum et animalium hactenus detecta. Prague: W. Gerle, 1777.

Sénebier, Jean. *Mémoires physico-chimiques, sur l'influence de la lumière solaire pour modifier les êtres des trois règnes de la nature, et surtout ceux du règne végétal.* 3 vols. Geneva: B. Chirol, 1782.

Sénebier, Jean. *Physiologie végétale contenant une description des organes des plantes et une exposition des phenomènes produits par leur organisation.* 5 vols. Geneva: J. J. Paschoud, 1799.

Smith, James Edward. "Observations sur l'irritabilité des végétaux." *Journal de physique* Rozier 33(1788):48–52.

Smith, James Edward. *An Introduction to Physiological and Systematical Botany.* London: Longman, Hurst, Rees, and Orme, 1807.

Tournefort, Joseph Pitton de. "Observations physiques touchant les muscles de certaines plantes." *Mémoires de l'Académie des Sciences,* Paris X(1669–1699):406–15.

Towson, Robert. "Objections against the Perceptivity of plants, so far as is connected by their external motion." *Tracts and Observations on the Natural History and Physiology.* London: The author, 1799.

Tupper, James Richard. *An Essay on the Probability of Sensation in Vegetables.* 2nd ed. London: R. and A. Taylor, 1818.

Uslar, Julius von. *Fragmente neuerer Pflanzenkünde.* Brunswick, 1794.

Whytt, Robert. *Essais physiologiques.* French trans. Paris: Les freres Estienne, 1759.

Whytt, Robert. *An Essay on the Vital and Other Involuntary Motions of Animals.* 2nd ed. Edinburgh: Hamilton, Balfour, and Neill, 1763.

Critical and Recent Works

Canguilhem, Georges. *La Connaissance de la vie.* 2nd ed. Paris: Vrin, 1965.

Guédès, Michel. "Augustin-Pyramus de Candolle et les mouvements végétaux." *Histoire de la nature* (1974), new series, fasc. 2.

Ors, Eugenio d'. *Du Baroque.* Paris: Gallimard, 1968.

Rousset, Jean. *La littérature de l'âge du baroque en France.* Corti, 1954.

Schiller, Joseph, and Tetty Schiller. *Henri du Trochet, 1776–1847.*

Le Matérialisme mécaniste et la physiologie générale. Paris: Blanchard, 1975.

Webster, Charles. "The Recognition of Plant Sensitivity by English Botanists in the Seventeenth Century." Isis 57(1966):5–23.

VI. PHILOSOPHY OF SCIENCE AND HISTORY OF IDEAS

Canguilhem, Georges. Etudes d'histoire et de philosophie des sciences. Paris: Vrin, 1968.

Dagognet, François. Le Catalogue de la vie. Paris: Presses Universitaires de France, 1970.

Daudin, Henri. De Linné à Jussieu. Méthodes de la classification et idée de serie en botanique et en zoologie (1740–1790). Paris: Alcan, 1926.

Foucault, Michel. Les Mots et les choses. Paris: Gallimard, 1966.

Foucault, Michel. L'Archéologie du savoir. Paris: Gallimard, 1969.

Foucault, Michel. L'Ordre du discours. Paris: Gallimard, 1971.

Jacob, François. La Logique du vivant. Paris: Gallimard, 1970.

Biographical Glossary

Adanson, Michel (1727–1806): French naturalist.

Alston, Charles (1685–1760): Scottish physician and botanist, professor at Edinburgh.

Aristotle (384–322 B.C.).

Aselli, Gaspard (1581–1626): Italian physician and ad anatomist.

Bachelot de la Pylaie, Jean-Marie (1786–1856): French naturalist and traveler.

Bacon, Francis (1561–1626): English philosopher and author.

Bayle, François (1622–1709); physician and professor at the University of Toulouse.

Bazin, Gilles-Auguste (?-1754): physician at Strasbourg, correspondent of the Academy of Sciences.

Bertholon, abbé Pierre (1742–1800): professor of physics at Montpellier.

Bichat, Xavier (1771–1802): anatomist and physiologist.

Birch, Thomas (1705–1766): English historian, clergyman, and editor of state papers.

Blair, Patrick (?–1728): physician and botanist, member of the Royal Society of London.

Boerhaave, Hermann (1668–1738): Dutch physician, professor of medicine and botany and of chemistry at Leyden.

Bondaroy, Auguste-Denis (1732–1789): member of the Academy of Sciences, nephew of Duhamel du Monceau.

Bonnet, Charles (1720–1793): Swiss philosopher and naturalist.

Borelli, Giovanni Alfonso (1608–1679): Italian physicist and astronomer, founder of the iatrophysical school.

Bose, Caspar (1703–1733): professor of botany at Leipzig.

Boyle, Robert (1627–1691): British physicist and chemist.

Bradley, Richard (1688–1732): physician and member of the Royal Society of London, professor of botany at Cambridge.

Broussonet, Pierre-Marie-Auguste (1761–1807): French naturalist, professor of botany at Montpellier.

Browallius, Johannes (1761–1807): bishop of Abro, botanist, and member of the Academy of Sciences of Stockholm.

Brown, John (1735–1788): Scottish physician.

Bruce, James (1730–1794): Scottish traveler and naturalist.

Bucquet, Jean-Baptiste (1746–1780): physician and professor of chemistry at the Faculty of Medicine, Paris.

Buffon, Georges Louis Leclerc, comte de (1707–1788): French naturalist, director of the Jardin du Roi and of the Royal Museum.

Bulliard, Jean-Baptiste (1752–1793): French botanist.

Burckhard, Johann Heinrich (1676–1738): German botanist, correspondent of Leibniz.

Camerarius, Rudolf Jakob (1665–1721): physician and botanist, director of the Botanical Gardens at Tübingen.

Camper, Pieter (1722–1789): Dutch anatomist and naturalist.

Candolle, Alphonse de (1806–1893): Swiss botanist, professor at Geneva.

Cardano, Girolamo (1501–1576): Italian mathematician, physician, and astrologer.

Carradori, Giovacchino (1758–1818): Italian naturalist, physician, and physicist.

Cavallo, Tiberius (1749–1809): Italian physicist, member of the Royal Society of London.

Cesalpinus, Andraeus (1519–1603). Italian botanist, philosopher, and physician, professor of materia medica and director of the botanical garden, Pisa.

Condorcet, Marie Jean Antoine de Caritat, marquis de (1743–1794): mathematician, philosopher, economist, and politician.

Correa de Serra, José Francisco (1721–1823): founder of the Academy of Sciences of Lisbon, member of the Royal Society of London.

Coulon, Julius Vitringa.

dal Covolo, Giambattista, conte.

Darwin, Erasmus (1731–1802): English physiologist and poet.

Dedu, Nicolas: French physician and botanist, lived at Montpellier during the second half of the seventeenth century.

Descartes, René (1596–1650): French scientist and philosopher.

Desfontaines, René-Louis (1750–1833): French botanist, member of the botanical section of the Institute.

Diderot, Denis (1713–1784): French encyclopedist and philosopher.

Dodart, Denis (1634–1707): French naturalist and physician, member of the Academy of Sciences of Paris.

Dodoens, Rembert (1517–1585): Dutch physician and botanist.

Dudley, Paul (1673–1751): English botanist.

Dufay, François de Cisternay (1698–1739): French chemist, member of the Academy of Sciences of Paris, first special director of the Jardin des Plantes.

Duhamel du Monceau, Henri-Louis (1700–1781): French engineer and agriculturist.

Duret, Claude (?–1611): President of the *presidial court* of Moulins.

Ellis, John (1710–1766): English naturalist, member of the Royal Society of London.

Farr, Samuel (1741–1795): English physician, studied at the University of Leyden.

Feldmann, Bernhard (1707–1777): Prussian physician, member of the Society of Observers of Nature of Berlin.

Fontenelle, Bernard le Bovier de (1657–1757): French writer of scientific popularizations.

Fouquet, Henri (1727–1806): French physician, professor of medicine at Montpellier.

Gaertner, Joseph (1732–1791): German botanist, professor of anatomy at Tübingen.

Galen (2nd century A.D.): Greek physician.

Gardeen, George (1649–1733): English botanist.

Geoffroy, Claude (1685–1752): French botanist and chemist, member of the Academy of Sciences of Paris.

Geoffroy, Etienne-François (1672–1731): French physician, botanist, and chemist, held the chair of medicine and pharmacy at the Collège de France.

Gerard, John (1545–1612): English botanist and barber-surgeon.

BIOGRAPHICAL GLOSSARY

Gersten, Christian Ludwig (1701–1762): German mathematician and botanist.

Gesner, Johannis (1709–1790): Swiss physician and mathematician, founder of the Physical Society of Zurich.

Girtanner, Christopher (1760–1800): German physician.

Gleditsch, Johann Gottlieb (1714–1786): German botanist.

Gleichen, Friedrich Wilhelm von (1717–1807): German naturalist.

Glisson, Francis (1597–1677): English physician, one of the founders of the Royal Society.

Gmelin, Johann Georg (1709–1755): German botanist, chemist, and explorer, professor at Tübingen.

Goethe, Johann Wolfgang von (1749–1832): German poet.

Gouan, Antoine (1733–1821): French physician and botanist, correspondent of Linnaeus and Rousseau.

Grew, Nehemiah (1641–1712): English plant physiologist.

Guettard, Jean Etienne (1672–1732).

Hales, Stephen (1677–1761): English chemist and naturalist.

Haller, Albrecht von (1708–1777): Swiss anatomist, physiologist, botanist, physician, and poet, professor at Göttingen.

Hartsoeker, Niklaas (1656–1725): Dutch physicist and histologist.

Harvey, William (1578–1657): English physician and anatomist, discoverer of the circulation of blood.

Hasselquist, Frederick (1722–1752): Swedish naturalist, student of Linnaeus.

Hedwig, Johann (1730–1799): German botanist, a founder of muscology.

Helmont, Jan Baptista van (1577–1644): Flemish physician and chemist.

Henschel, August Wilhelm (1790–1856): German botanist.

Henshaw, Thomas (1618–1700): member of the Royal Society of London.

Hertel, Johann Gottlob.

Hill, John (1716–1775).

Hippocrates (460–377 B.C.): Greek physician.

Hoffmann, Friedrich (1660–1742): German physician.

Home, Henry, Lord Kames (1696–1782): Scottish lawyer and philosopher.

Hooke, Robert (1635–1703): English experimental philosopher.

Hooper, Robert (1773–1835): English botanist.

Humboldt, Alexander von (1769–1859): German naturalist, traveler, and statesman.

Hunter, John (1728–1793): British anatomist and surgeon.

Huygens, Christian (1629–1695): Dutch physicist and astronomer.

Jallabert, Jean (1712–1768): Swiss physicist and mathematician.

Jaucourt, Louis de (1704–1779): wrote articles on botany, physics, and medicine for the *Encyclopédie*.

Jussieu, Antoine de (1686–1758): French physician and botanist, professor at the Jardin des Plantes.

Jussieu, Bernard de (1699–1777): brother of Antoine, laid foundation for a natural system of plant classification.

Jussieu, Antoine-Laurent de (1748–1836): French botanist, professor at the Jardin des Plantes, elaborated his uncle Bernard's system of classification.

Kant, Immanuel (1724–1804): German metaphysician and transcendental philosopher.

Kastner, Karl Wilhelm (1783–1857): German botanist.

Keill, John (1671–1721): Scottish mathematician and astronomer.

Knight, Thomas Andrew (1759–1838).

Koelreuter, Josef Gottlieb (1733–1806): German botanist, pioneer in hybridization experiments with plants.

La Brosse, Guy de (1586–1641): French botanist, first director of the Jardin des Plantes.

La Hire, Philippe de (1640–1718): French astronomer and mathematician.

Lamarck, Jean-Baptiste de (1744–1829): French naturalist and evolutionary philosopher.

La Metherie, Jean Claude de (1743–1807): French physicist and naturalist, adjunct professor of natural history at the Collège de France.

La Mettrie, Julien Offroy de (1709–1751): French physician and materialistic philosopher.

Lamouroux, Jean (1779–1825): French naturalist with a particular interest in the natural history of the sea.

Lamy, Bernard (1640–1715): French ecclesiastic and scholar.

BIOGRAPHICAL GLOSSARY

La Quintinie, Jean de (1626–1688): agronomist, responsible for the gardens at Versailles.

Leeuwenhoek, Anton van (1632–1723): Dutch naturalist and microscopist.

Leibniz, Gottfried Wilhelm von (1646–1716): German philosopher and mathematician.

Lindsay, John (of Westmorland, Jamaica, before 1800).

Linnaeus, Carolus (1707–1778): Swedish botanist, a founder of modern systematic botany.

Logan, James (1674–1751): American statesman and amateur botanist, correspondent of Sloane and Peter Collinson.

Ludwig, Christian Gottlieb (1709–1773): German botanist and physician.

Magnol, Pierre (1638–1715): French physician and botanist, professor at the Royal Gardens of Montpellier, originated classification of plants by families.

Mairan, Jean-Jacques Dortous de (1678–1771).

Major, Jean Daniel (1639–1693): physician and botanist, founder of the Botanical Gardens of Kiel.

Malpighi, Marcello (1628–1694): Italian physician and anatomist.

Mariotte, Edme (1620–1684): French physicist.

Martí, Antonio de (1750–1832): Spanish botanist.

Maupertuis, Pierre Louis Moreau de (1698–1759): French mathematician and astronomer.

Mayow, John (1645–1679): English physiologist and chemist.

Micheli, Pierre-Antoine (1679–1737): botanist attached to P. Boccone, botanist of the Grand Duke of Tuscany.

Miller, Philip (1691–1771): student of John Ray, superintendent of the Garden of Apothecaries at Chelsea.

Moller, Hans (1736–1796): German botanist.

Montesquieu, Charles de Secondat (1689–1755): French lawyer, man of letters, and political philosopher.

Morland, Samuel (1625–1695): English diplomat and inventor.

Musschenbroeck, Pietr van (1692–1761): Dutch mathematician and physicist.

Mustel, Nicolas Alexander.

Necker, Noël Joseph de (1730–1793): French botanist and physician.

BIOGRAPHICAL GLOSSARY

Needham, John Turberville (1713–1781): English naturalist.

Newton, Isaac (1642–1727): English natural philosopher and mathematician.

Nieuwentyt, Bernard (1654–1718).

Palisot de Beauvois, Ambroise (1752–1820): correspondent of the Royal Academy, member of the Botanical Section of the Institute.

Parent, Antoine (1666–1716): scholar, mathematician, member of the Academy of Sciences.

Parsons, James (1705–1770): English physician, member of the Royal Society of London.

Pecquet, Jean (1622–1674): French physician and anatomist.

Percival, Thomas (1740–1804): English physician, member of the Royal Society of London.

Perrault, Claude (1613–1688): French architect, physician, and physicist.

Peschier, Jacques-Louis (1769–1832): Geneva botanist.

Plato (427–347 B.C.): Greek philosopher.

Pontedera, Giulio (1688–1757): Italian botanist, professor at Padua.

Porta, Giambattista della (1538–1615): Italian physicist.

Pultney, Richard (1730–1801): English physician and botanist.

Ray, John (1627–1705): English naturalist.

Réaumur, René Antoine Ferchault de (1683–1757): French physicist and naturalist.

Redi, Francesco (1626–1697): Italian physician, naturalist, and poet.

Reichel, George Christian (1724–1755): German mineralogist and botanist.

Renéaume, René-Louis (1675–1739): French botanist, member of the Academy of Sciences.

Reynier, Louis (1762–1824): French naturalist and agronomist.

Robinet, Jean-Baptiste (1735–1820).

Rotheram, John (1750–1804): English disciple of Linnaeus.

Rousseau, Jean-Jacques (1712–1778): French philosopher and author.

Sarrabat, Nicolas (pseudonym "de le Baïsse," 1698–1739): Jesuit, professor of mathematics, won three prizes from the Bordeaux Academy.

Saussure, Horace Bénédict de (1740–1799): Swiss naturalist, professor at Geneva.

Schelver, Friedrich Joseph (1778–1832): German botanist, professor at Heidelberg.

Scopoli, Giovanni Antonio (1723–1788): Italian naturalist, member of the Society of Nature Lovers and of the Societies of Science of Padua, Naples, and Turin.

Sénebier, Jean (1742–1809).

Siegesbeck, Johann Georg (1686–1755): professor of botany at St. Petersburg.

Smellie, William (1745–1795): Scottish printer and antiquary, member of the Royal Society of Edinburgh.

Smith, James Edward (1759–1828).

Spallanzani, Lazarro (1729–1799): Italian biologist, professor at Pavia.

Sprengel, Christian Konrad (1750–1816): German botanist.

Stillingfleet, Benjamin (1702–1771): English botanist, introduced the sexual system to England.

Swammerdam, Jan (1637–1680): Dutch naturalist.

Tournefort, Joseph Pitton de (1656–1708): French botanist and traveler, one of the founders of modern systematic botany.

Towson, Robert (before 1800).

Trembley, Abraham (1700–1784): Swiss naturalist.

Tull, Jethro (1674–1741): English agriculturist.

Tupper, James Perchard (17?–18?).

Uslar, Johann Julius von (1762–1838).

Vaillant, Sébastien (1669–1722): French botanist.

van Marum, Martinus (1750–1837).

Vicq d'Azir, Félix (1748–1794): French physician.

Volta, Giovanni Serafino (second half of the eighteenth century).

Walker, John (1731–1804).

Wharton, Thomas (1614–1673): English physician.

Whytt, Robert (1714–1766).

Wolf, Christian von (1679–1754): German philosopher and mathematician.

Woodward, John (1665–1728): English physician and naturalist.

Name Index

Adanson, Michel, 107, 120
Alston, Charles, 92, 119, 120, 122, 139, 147
Amici, Giovanni Battista, 189, 190
Aristotle, 19, 20, 73, 152
Aselli, Gaspard, 31

Bachelot de la Pylaie, Jean-Marie, 219n.77
Bacon, Francis, 220n.11
Bayle, François, 202n.28
Bazin, Gilles-Auguste, 44, 45
Bell, Charles, 225n.23
Bertholon, abbé Pierre, 47
Bichat, Xavier, 189, 191
Birch, Thomas, 204n.15
Blair, Patrick, 97
Boerhaave, Hermann, 80–89, 176, 178
Bondaroy, Auguste-Denis, 123
Bonnet, Charles, 47, 63, 117, 120, 124, 125, 128, 129, 139, 145, 150, 153–155, 164–168, 176, 177
Borelli, Giovanni Alfonso, 53
Bose, Caspar, 223n.2
Boyle, Robert, 209n.83
Bradley, Richard, 70, 73–75, 92, 96, 99, 193, 218n.56
Broussonet, Pierre-Marie-Auguste, 173, 229n.56

Browallius, Johannes, 222n.119
Brown, John, 157
Brown, Robert, 189
Bruce, James, 225n.23
Bucquet, Jean-Baptiste, 208n.70
Buffon, Georges Louis Leclerc, comte de, 51, 80–89, 135
Bulliard, Jean-Baptiste, 113
Burckhard, Johann Heinrich, 214n.23

Camerarius, Rudolf Jakob, 3, 92, 95, 103, 120
Camper, Pieter, 230n.68
Candolle, Alphonse de, 168, 169
Canguilhem, Georges, 202n.30
Cardano, Girolamo, 14
Carradori, Giovacchino, 224n.11
Cavallo, Tiberius, 225n.16
Cesalpinus, Andraeus, 14, 20, 120, 193
Comte, Auguste, 16
Condorcet, Marie Jean Antoine de Caritat, marquis de, 136
Correa de Serra, José Francisco, 114

Coulon, Julius Vitringa, 156
Courtes, Francis, 211n.99
Cudworth, Ralph, 78
Cuvier, Georges, 50

Dagognet, François, 21
dal Covolo, Giambattista, conte, 152, 184, 185
Darwin, Erasmus, 128, 150, 157, 158
Daudin, Henri, 212n.113
Dedu, Nicolas, 203n.6
Descartes, René, 22, 76, 85, 191
Desfontaines, René-Louis, 129
Diderot, Denis, 83, 84
Dodart, Denis, 165
Dodoens, Rembert, 201n.26
Dudley, Paul, 101
Dufay, François de Cisternay, 227n.39
Duhamel du Monceau, Henri-Louis, 18, 44, 49, 51, 64–70, 120, 164–172
Duret, Claude, 21
Dutrochet, Henri, 189–191

Ellis, John, 230n.75
Empedocles, 93

Farr, Samuel, 223n1.
Feldmann, Bernhard, 17, 18
Fontenelle, Bernard le Bovier de, 217n.49
Foucault, Michel, 6, 187
Fouquet, Henri, 209n.73

Gaertner, Joseph, 219n.76
Galen, 20–21
Galien, Claude, 20, 21
Gardeen, George, 217n.49
Geoffroy, Claude, 102
Geoffroy, Etienne-François, 43, 44
Gerard, John, 202n.27
Gesner, Johannis, 48

Girtanner, Christopher, 157, 158
Gleditsch, Johann Gottlieb, 104, 105, 107
Gleichen, Friedrich Wilhelm von, 105, 109, 140
Glisson, Francis, 36
Gmelin, Johann Georg, 153, 154
Goethe, Johann Wolfgang von, 223n.124
Grew, Nehemiah, 10, 17, 18, 27, 37, 39, 41, 42, 64, 66–68, 116, 120
Guettard, Jean Etienne, 30, 49, 59, 60, 61, 65, 113

Hales, Stephen, 2, 30, 51, 53–67, 80, 85, 88
Haller, Albrecht von, 24, 117, 151, 152, 153, 174
Hartley, David, 164
Hartsoeker, Niklaas, 115, 116
Harvey, William, 2, 31, 34, 35, 38
Hasselquist, Frederick, 215n.26
Hedwig, Johann, 112
Helmont, Jan Baptista van, 73
Henschel, August Wilhelm, 123
Henshaw, Thomas, 204n.15
Herodotus, 21
Hertel, Johann Gottlob, 59
Hill, John, 164, 166–168
Hippocrates, 30
Hoffmann, Friedrich, 85, 86, 152, 178
Hofmeister, Wilhelm, 190
Home, Henry, Lord Kames, 226n.33
Hooke, Robert, 27, 208n.65
Hooper, Robert, 223n.2
Humboldt, Alexander von, 225n17, 226n.28
Hunter, John, 4, 168, 169

INDEX

Huygens, Christian, 210n.93

Ingenhousz, Jean, 190

Jacob, François, 77, 110, 135
Jaucourt, Louis de, 30
Jussieu, Antoine de, 120, 193
Jussieu, Antoine-Laurent de, 210n.90
Jussieu, Bernard de, 111

Kant, Immanuel, 85, 86, 89
Kastner, Karl Wilhelm, 218n.56
Knight, Thomas Andrew, 4, 150, 169, 173
Koelreuter, Josef Gottlieb, 129–131, 135

La Hire, Philippe de, 165
Lamarck, Jean-Baptiste de, 171, 174, 184, 185
La Metherie, Jean Claude de, 205n.31, 225n.23
La Mettrie, Julien Offroy de, 22, 82–84, 89
Lamouroux, Jean, 217n.48
La Quintinie, Jean de, 212n.117
Lavoisier, Antoine, 190, 191
Leeuwenhoek, Anton van, 102
Leibniz, Gottfried Wilhelm von, 146
Limoges, Camille, 216n.33
Lindsay, John, 4, 150, 172, 173, 184, 185
Linnaeus, Carolus, 17, 92, 103, 112, 126, 128, 134, 139, 140, 166–167, 183
Logan, James, 104, 105
Ludwig, Christian Gottlieb, 216n.26

Magnol, Pierre, 207n.57
Mairan, Jean-Jacques Dortous de, 166

Major, Jean Daniel, 2, 31, 32
Malpighi, Marcello, 9, 10, 16–18, 27, 39–47, 51, 66, 67, 115, 116
Mariotte, Edme, 31, 33, 37, 71–73, 78, 79
Martí, Antonio de, 218n.56
Maupertuis, Pierre Louis Moreau de, 114, 135
Mayow, John, 66
Micheli, Pierre-Antoine, 111
Miller, Philip, 96
Mirbel, C.-F. Brisseau de, 189
Mohl, Hugo von, 190, 191
Möller, Hans, 119, 218n.56
Montesquieu, Charles de Secondat, 210n.96, 211n.98
Morland, Samuel, 120
Musschenbroeck, Pietr van, 205n.35
Mustel, Nicolas Alexander, 66, 75

Necker, Noël Joseph de, 125, 126
Needham, John Turberville, 109
Nieuwentyt, Bernard, 208n.66

Olby, Robert, 221n.96, 222n.116
Ors, Eugenio d', 183

Palisot de Beauvois, Ambroise, 216n.41
Parent, Antoine, 62
Parsons, James, 70, 73, 74
Pecquet, Jean, 31
Percival, Thomas, 226n.26
Perrault, Claude, 2, 31–37, 39, 44, 65
Peschier, Jacques-Louis, 179
Pliny, 21, 86
Pontedera, Giulio, 222n.108
Porta, Giambattista della, 197
Priestley, Joseph, 190

Pringsheim, Nathaniel, 190
Pultney, Richard, 222n.116

Ray, John, 78, 120
Réaumur, René Antoine Ferchault de, 47, 112
Redi, Francesco, 16
Reichel, George Christian, 207n.58
Renéaume, René-Louis, 49, 62
Reynier, Louis, 119
Ritterbush, Philip, 221n.106
Robinet, Jean-Baptiste, 176, 177
Rotheram, John, 222n.115
Rousseau, Jean-Jacques, 12, 140

Sarrabat, Nicolas, 62, 63, 69, 70
Saussure, Horace Bénédict de, 48, 49
Schelver, Friedrich Joseph, 123
Schleiden, M.-J., 189, 190
Schwann, Théodore, 189
Scopoli, Giovanni Antonio, 174
Sénebier, Jean, 9, 168, 190
Siegesbeck, Johann Georg, 144
Smellie, William, 124
Smith, James Edward, 152–154
Spallanzani, Lazzaro, 3, 92, 117–125, 147
Sprengel, Christian Konrad, 129–134, 189
Stillingfleet, Benjamin, 223n.119
Swammerdam, Jan, 97

Theophrastus, 86
Thuret, Gustave Adolphe, 190
Tournefort, Joseph Pitton de, 92, 116, 117, 120

Towson, Robert, 226n.29
Trembley, Abraham, 222n.118
Tull, Jethro, 45, 46, 71, 73–75
Tupper, James Perchard, 225n.17, 226n25.

Uslar, Johann Julius von, 225n.16

Vaillant, Sébastien, 98, 102, 120
van Marum, Martinus, 154–158, 184, 185
Vicq d'Azir, Félix, 212n.113
Volta, Giovanni Serafino, 119

Walker, John, 208n.64
Wharton, Thomas, 36
Whytt, Robert, 171
Wolf, Christian, 208n.66
Woodward, John, 56, 57

INDEX